MULTIFAMILY HOUSING
集合住宅

MULTIFAMILY HOUSING
集合住宅

[澳] 汉娜·詹金斯 编著

齐梦涵 译

广西师范大学出版社
·桂林·

images Publishing

图书在版编目(CIP)数据

集合住宅/(澳)汉娜·詹金斯编著;齐梦涵译. —桂林:广西师范大学出版社,2018.1
ISBN 978－7－5598－0001－5

Ⅰ.①集… Ⅱ.①汉… ②齐… Ⅲ.①集合住宅－建筑设计 Ⅳ.①TU241.2

中国版本图书馆 CIP 数据核字(2017)第 295965 号

出 品 人:刘广汉
责任编辑:肖　莉
助理编辑:齐梦涵
版式设计:吴　迪
广西师范大学出版社出版发行
(广西桂林市五里店路9号　　邮政编码:541004)
(网址:http://www.bbtpress.com)
出版人:张艺兵
全国新华书店经销
销售热线:021－31260822－882/883
恒美印务(广州)有限公司印刷
(广州市南沙区环市大道南路334号　邮政编码:511458)
开本:635mm×965mm　　1/8
印张:30　　　　　　字数:30 千字
2018 年 1 月第 1 版　　2018 年 1 月第 1 次印刷
定价:256.00 元

如发现印装质量问题,影响阅读,请与印刷单位联系调换。

目录

6　前言

可持续住宅

14　大河被动式节能屋
　　Steinsvik建筑事务所

21　环绕森林别墅
　　海伦&哈德建筑事务所

27　GI多家庭住宅
　　Burnazzi Feltrin建筑事务所

33　萨隆根35号
　　Kjellgren Kaminsky建筑事务所

39　贝尔斯联排房屋
　　基兰·汀布莱克

45　扬巷住宅
　　贾斯汀·马利亚

51　炉窑公寓
　　GBD建筑事务所

经济适用公寓

59　城市中心公益住房
　　LAPS建筑事务所与MAB建筑事务所

65　普莱因·苏蕾公寓
　　rh+建筑事务所

71　理查森公寓
　　大卫·贝克建筑事务所

77　百老汇住房
　　凯文·戴利建筑事务所

83　北帕克公寓
　　乔纳森·西格尔，美国建筑师协会会员

89　瓦尼卡斯公益住房
　　吉耶尔莫·巴斯克斯·孔苏埃格拉建筑事务所

95　里士满街60号
　　蒂普尔建筑事务所

101　黑白双子公寓
　　卡萨诺瓦事务所与赫尔南德斯建筑事务所

豪华住宅

109　18号公寓
　　Aytac建筑事务所

115　维特拉公寓
　　里伯斯金工作室

121　镜屋
　　彼得·毕希勒建筑事务所

127　白石工作室
　　本杰明·哈尔设计公司

133　锡尔特公寓
　　比罗^普罗伯茨建筑事务所

139　318号住所
　　艾伦伯格·弗雷泽

145　威尔士街公寓
　　MA建筑事务所与Neometro建筑事务所

创意社区

153　马歇尔巷公寓
　　斯泰普尔顿·艾略特设计公司

159　NOIE集合住宅
　　YUUA建筑设计事务所

165　狼溪住宅
　　麦卡曼特&达雷特建筑事务所

171　顶点公寓
　　Rothelowman设计公司

177　新月公寓
　　Hyla事务所

183　桶架公寓
　　贝兹斯玛特设计公司

创新设计

191　科尔多瓦城郊住宅
　　卡达瓦尔和索拉-莫拉莱斯设计公司

197　立方体公寓
　　霍金斯\布朗建筑事务所

203　狄龙公寓
　　史密斯-米勒建筑事务所与霍金森建筑事务所

209　阿尔法公寓
　　托尼·欧文公司

215　汉普顿环道公寓与联排别墅
　　柯林斯·加达耶建筑事务所

221　布里克社区
　　Dekleva Gregorič建筑事务所

227　上院公寓
　　杰克逊·克莱门茨·巴罗斯建筑事务所

233　雕刻公寓
　　A-lab设计公司

238　索引

前言
阿维·弗里德曼

多户住宅：历史、挑战与机遇

人类进入21世纪，社会上发生的新挑战和新机遇如暴风般席卷而来，影响了世界上大多数国家的人们在接下来的若干年中对住房条件的选择。从城市层面来说，一些国家对停止城市面积扩张的需求和另一些国家中迅速的城市化进程使城市密度不断增加。人们要求城市为更多市民提供住房，同时还希望降低碳的排放量，减少城市对生态的破坏，这自然使集合住宅受到人们的欢迎和重视。这篇介绍性的文字概述了集合住宅的发展历史，还说明了近期社会转变的情况以其对建筑设计的影响。

回首往昔

纵观历史，人类曾为采集食物、获得安全和社会地位而聚集起来。定居点（农业或商贸）为人类文明的发展做出过极大贡献——人类交换想法和信息的行为推动了技术和文化的发展，而这种行为通常发生在人类聚居的社区之中。了解与学习城市进化过程中多家庭共生的一些标志性案例并从中获得经验与知识是非常有价值的。

大约6世纪开始，城市的功能开始与传统的商业和人口中心发生分离，这种变化在古代文明中十分普遍。罗马帝国覆灭之后，掠夺者袭击了城镇，洗劫了村庄，迫使人口迁移到偏远的地区。

另外一些人决定留在城市，并定居于修筑了防御工事的竞技场废墟中，法国阿尔勒的情况就是如此，这种做法强化了墙是保护人民最好的方式这一观念。城墙修建好之后，城市的边界就被固定下来，墙壁之内的城市发展则必然被压缩成密集的形态，这种城市形态是人口增长的结果。

几个世纪之后，中世纪社会普遍认为家庭生活与工作是互补的，工商业活动与家庭生活相同，也都发生在室内。因此随着商业活动的重要程度日益加大，有临街空地的房屋也被认为更有价值。其结果便是建筑平面图被设计得狭窄而纵深，一般来说，建筑物的宽度与进深比为1∶6。即便是在较大的中世纪城市中，墙壁外围到市中心的距离也不会超过500米，这种相对较小的面积只能靠高密度来弥补。这种压缩了的城市规模不仅仅是巍然不动的要塞城墙造成的，徒步旅行和自给自足的需求也决定了城市规模不可能过大。

中世纪时期人们对密集城市形态所做的创新发展的一个案例便是英格兰的切斯特市。切斯特市之所以如此特别，是因为那里有著名的"购物长廊"，这是一种狭长的木制多用途住宅式建筑，高高地架设在街道之上。位于上层的带有居住场所的店铺，由其外面架高的长廊连通起来，长廊下方是便于人流移动的街道。遗憾的是，切斯特的许多创新城市概念都在文艺复兴时期遗失了。

文艺复兴运动包含了大量城市改造工程，建筑师们试图通过新开发项目和建筑修复，来恢复城市往日古典文明的荣光。中世纪人口规模的城市逐渐被令人印象深刻、规模更大的城市所取代。无论是宽阔的林荫大道，还是一览无余的美丽风景，抑或是熙熙攘攘的市民广场、笔直的街区布局，都是符合这个时期的设计指导原则的。在城市中，小型工商业从住宅中脱离出来，搬到独立的建筑物中。家庭和工作的分离改变了社区的关系，结果众多城市因经济和艺术带来的财富而繁荣，也因社区的贫困而步履维艰。

工业革命之后，有效的施工技术改变了许多多户住宅的构建。那个时代的早期，设计师们仍然对公众的认知很敏感。为了确保销售及出租房屋的收益，有吸引力的细节——例如凸窗和升高的入口——被加入到多户住宅的建筑设计中。然而，随着时间的推移，这种住宅的需求量急剧增加，建筑者也随之出售品质低一些的住宅——进深更短、宽度更窄的房屋顺应这种趋势逐渐发展起来。过度拥挤的住房条件在所有的城市中缓慢而小范围地发生——19 世纪工人居住的单调乏味的住宅一般被限制在城市的某一特定的区域，因为这种连续的结构具有独立自主、能为居住于其中的家庭提供一定程度的隐私等优点，所以相对更受到居住者的青睐。不过，这种多户住宅也因为三面墙壁与相邻单元连接，只在入口一侧开有窗户而缺少对流通风。

北美的城市化发展模式也导致了低收入家庭的生活条件低下。费城是采用英式住宅概念的第一座北美城市，不过这里更高的土地价值与棋盘式的街道布局创造出了多达 10 到 12 个居住单元的街区。随着北美的工业进步，多户住宅变得越来越小，例如，两块 625 平方英尺（58 平方米）的土地被分成若干个三室加一间"暗室"（指面积是普通房间的 1/3 并且没有窗户的房间）的住房，这种情况相当普遍。越来越多的农村人口涌入城市，导致低收入工人与移民开始面临住房短缺的问题，传统中产阶级的多户住宅也因此向廉价出租屋转变。人们自古以来一直认为密集的城市社区是保护市民的安全港湾，这种观念到了这时也开始改变，越来越多的人们认为城市是一个充满污染、疾病与贫穷的地方。人们对乡村生活的看法也随之改变，曾经被认为不适合居住的地方，现在成了人们想要生活的地方。人们开始渴望开放的空间、大自然，以及健康的生活环境，当然他们还要去市区上班，所以不能生活得离城市太远。另外，人们想要逃离的不单是城市的污染和贫困，还有那里的过度拥挤，未开发的乡村正好为人们对宽敞生活环境的追求提供了条件，这导致了城市郊区的诞生。

现代挑战和机遇

现代社会越来越意识到其发展是以破坏环境为代价的，人们开始重新审视常见的住宅设计与施工实践，希望能够减少其对环境的影响。大型独立住宅的不断增加与其诸多的负面影响证明了区域性的活动会造成全球性的影响。人们也意识到如果想要扭转这些活动造成的影响，必须尽快改变过去的做法。

这些改变的核心是人类对地球资源的再生速度小于其消耗速度的认知。人们对全球变暖的负面影响提高了警觉，并且需要对其做出回应，这在某种程度上改变了设计师考虑建筑设计时的首要任务，使他们更加支持多户住房的概念。城市规划师认为建设能够容纳多个家庭单位且具备多种用途的高密度可持续性社区或建筑对公共交通系统极为重要，这能够降低私家车的使用率，减少二氧化碳的排放量。

减少对化石燃料、淡水等不可再生自然资源的消耗成为趋势，并引领了高能效建筑、零能源设计、被动式太阳能获取策略、环保发电系统的发展，其产物包括光伏板、水收集和循环利用等。这种趋势也推动了绿色屋顶，由回收材料制作的物品等产品的引进与使用。

经济衰退对社会造成了严重的损害，影响了全球市场及公民的个人生活。找一份工作并且获得稳定的收入变得不再像过去那样普遍，因而对于首次购房的人们来说，想要在人口密集的市中心购买住房是件非常困难的事情。另外，人们负担能力的差距逐渐加大，房价上涨速度也随之超过了家庭收入的增长速度。

在过去，购房者总是寻找更大的住宅，这经常使他们的财务状况捉襟见肘，随着2008年爆发的全球性房地产泡沫，一种不同的态度开始占据主导地位。越来越多的潜在购房者不再能够承担独立住宅的负担，其结果便是人们对更加便宜、面积更小、更节能的集合住宅的需求开始加大。微单元与灵活的室内空间等集合住宅项目也开始出现。而在室内方面，设计师不断开发新方法，使整合各种公共设施的工作可以与建筑施工过程同步开展，有效降低其建造和使用成本。为了节约时间，降低成本，全球的建筑行业也普遍接受了使用预制房屋构件建造房屋。

近期社会人口结构的变化影响着人们的生活，也影响着人们居住的房屋。第二次世界大战之后的传统家庭由养家糊口的父亲、家庭主妇与受到照料的孩子所组成，这个模式是如此深入人心，以至于住宅建筑商可以简单地把大多数潜在客户当作同类型的购买者来对待。而目前快速变化的人口构成与生活方式的新趋势则要求当代住宅类型向小型、灵活、高效转变。

人口构成变化中最主要的变化是非传统家庭与小型家庭数量的增加，他们通常更愿意在集合住宅中寻找住所。独身家庭、丁克家庭与单亲家庭的数量在过去50年里增长了好几倍，它要求设计师重新考虑当前最普遍的住宅原型。这些住户为了靠近工作地点、便利设施和公共交通，通常选择在市中心寻找住房。多户住宅——因其混合的单位类型与多样化的购买者——正是这些群体所寻求的，因此该种建筑类型的价值也不断上升。

政府的决策制定者、建筑师和整个房屋建筑业都不得不考虑一个事实，那便是在很多国家中老龄人口都将在总人口中占有很大比例。所有老年人都能在养老辅助生活机构中解决住房问题的可能性很小，所以许多老年人很有可能需要在自己的家中养老，而在这之中，有些公寓需要被特殊改造，以使老年人能够尽可能长久地生活于其中。另一种可能会增加的建筑类型是为多代同堂的家庭建造的集合住宅，这种建筑类型要在同一座建筑物内部为所有家庭成员提供私人空间和公共空间。

数字技术的飞速发展也影响着集合住宅的设计。远程办公及生活工作一体化的住宅设计因其在经济、环境和社会方面所具有的优势而逐渐兴起并大受欢迎，人们通过居家办公不单能减少与工作相关的支出，还能减少上下班开车的时间，获得更多的休闲时间。在未来的几年里，对兼具生活办公两种功能的住宅的需求还会进一步增长。

纵观历史，集合住宅的设计师显示出了他们能根据具体环境不断调整这种住房形式的非凡能力。我们的时代也不例外，当代社会面临的挑战要求建筑师们必须不断创新，未来的新挑战很可能也会引领其他设计方法的出现。本书概述的趋势和其产生的建筑形式正是集合住宅的设计与建造在未来还将继续发展。

创造一个社区
Creating a Community

可持续住宅

SUSTAINABLE HOUSING

Steinsvik建筑事务所

大河被动式节能屋

STEINSVIK ARCHITECTURE

PASSIVHUS STORELVA

Location Trosmø, Norway **Units** 7 **Area** 12,831 ft² (1192 m²) **Photography** Ravn Steinsvik

Storelva is the third passive house project from Steinsvik Arkitektkontor. Located in the city of Trosmø, Norway, just above the Arctic Circle, the project has a primary focus on wood technology. Due to Norwegian environmental elements, low-energy construction techniques based on established technologies, which use light insulated timbered construction with indoor moisture barrier and outdoor windbreaker/moisture protection, were developed. However, researchers later concluded that passive housing using these technologies was not suited to the Norwegian climate due to the danger condensation posed, causing subsequent mold formation in the insulation layer. The Storelva architects saw an opportunity to revert to a diffusion-open construction, with the inside of the structure bearing masses of wood, and the outdoor made moisture resistant through use of mineral wool insulation. This meant the insulation thickness would be equal-to or greater-than the thickness of the wood elements. Their calculations showed that the condensation would occur in the insulation part of the wall, and thus not damage the construction or affect the in-building environment.

PASSIVHUS STORELVA

The seven compact houses are divided over three levels and sit on 322-square-foot (30-square-meter) ground planes, with a superstructure for stairs leading to the roof terraces. In order to achieve passive housing status, the following technical devices were included: geothermal collectors, sun collectors, large south-facing glass windows with low heat transmission, heat recycling in the ventilation system, a heat pump for air and water, and airtight outer-wall construction.

The basis for the construction principle was to use natural materials wherever possible, and as few chemical additives as possible to reduce harmful substances being released, both inside and into nature. Therefore, no form of surface treatment has been used on any construction elements. The massive wood components are glued together with low-emitting heat tempered glue. The whole construction is screwed together and can be taken apart with a screwdriver if moving or modifying is desired.

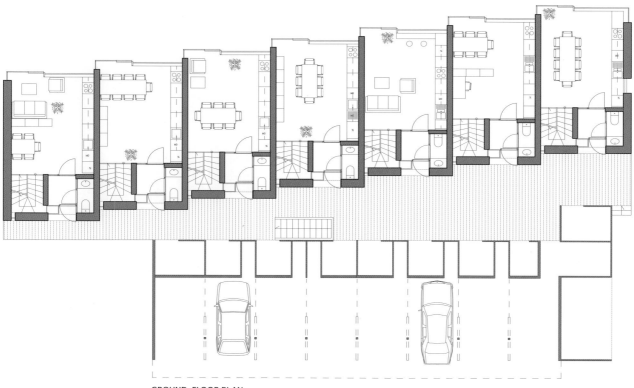

GROUND-FLOOR PLAN

海伦&哈德建筑事务所
环绕森林别墅

HELEN & HARD
RUNDESKOGEN

Location Norway **Units** 113 **Area** 646–1507 ft² (60–140 m²) **Photography** Emile Ashley

Rundeskogen is situated in an infrastructural node between three city centers on the west coast of Norway. Single-family houses and small-scale housing projects dominate the region, creating a context that accentuates the exceptional height and volume of the structure. The density and concentration of the project was developed to keep a required distance from a recently discovered Viking grave on the neighboring hillside.

The three towers contain 113 units in total, ranging from 646 square feet (60 square meters) to 1507 square feet (140 square meters), with the highest tower reaching 16 stories. The core construction is concrete, while secondary parts are made from timber framework. Emphasis has been placed on balancing the tall building typology by the generous and attractive public green spaces on the ground. To minimize the footprint of the three towers and retain the neighbors' view of the fjord, the first apartment floors have been lifted off the ground, cantilevering from the core to create outdoor spaces on the ground level for play and recreation.

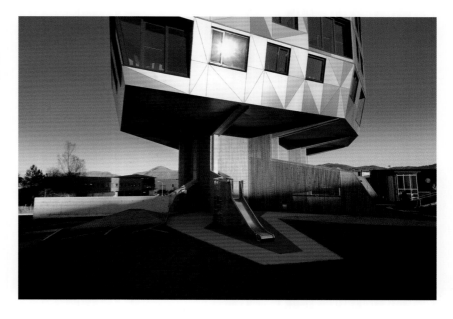

Internally, each apartment has an integrated winter garden, which has fully insulated glass façades allowing for flexible, year-round use. Other environmental features include solar collectors on the roof, heat recovery from graywater, and ground source heat pumps.

The organizing element of the entire project is a star-shaped core structure of concrete, in which fins are extended as separation walls between the flats. The prismatic shape of the plan is derived from optimizing the floor plans according to views and sun, as well as the desire to provide diagonal views in between and around the towers. This layout provides special volumetric qualities, which allow light and shade to gradually shift around the façades, an effect that is further emphasized by triangular panels, which reflect the light differently as one moves past the building. On the ground floor, the fins and bracing elements of the stem-like core spread out as roots and integrate social meeting places.

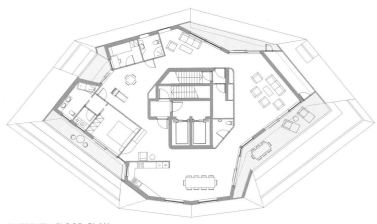
SIXTEENTH-FLOOR-PLAN

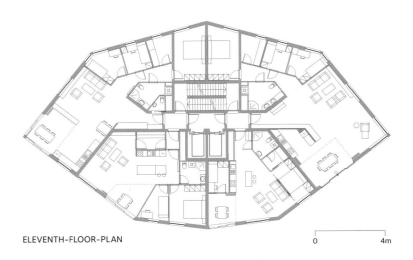
ELEVENTH-FLOOR-PLAN

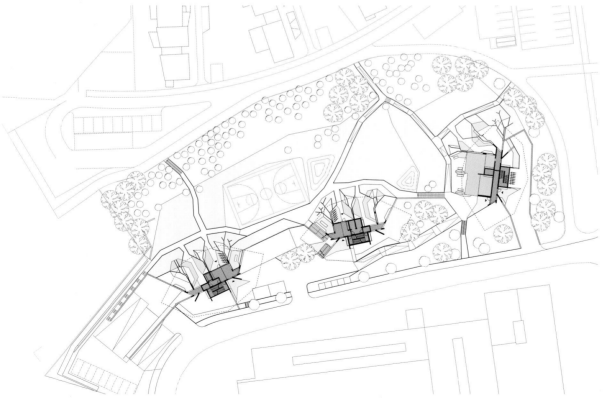

SITE PLAN

Burnazzi Feltrin建筑事务所

GI多家庭住宅

BURNAZZI FELTRIN ARCHITECTS

GI MULTI-FAMILY HOUSING

Location Trento, Italy **Associates/contributors** Pegoretti Paolo **Units** 3 **Area** 3660 ft² (340 m²) **Photography** Carlo Baroni

The GI multi-family housing is located on a slope in Ischia, a small town to the southeast of Pergine Valsugana. The building consists of three apartments, one on each floor, with the last floor designed as a duplex. Large windows allow for a beautiful view of Lake Caldonazzo and take advantage of the excellent year-round light. The peculiarity of the location and its existing buildings inspired the architects to make a very compact and uniform construction volume. Their aim was to reduce the visual impact of the main front—the one visible from the lake—by using loggias to create continuity between the front and the pitched roofs. The roof doesn't jut out from the walls and is actually integrated into the volume, which sees the external walls plastered with colors inspired by the local natural soils, whereas the balconies and terraces have a transparent parapet. The niches of the loggias, a reinterpretation of the historic buildings in the country, along with the lateral walls outside the window frames, sliding doors and cantilevered slabs, are dark colored.

GI MULTI-FAMILY HOUSING

SITE PLAN

GI MULTI-FAMILY HOUSING 31

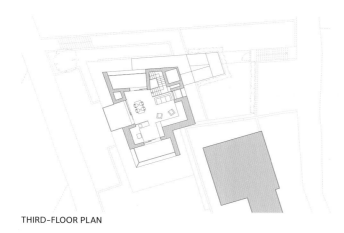
THIRD-FLOOR PLAN

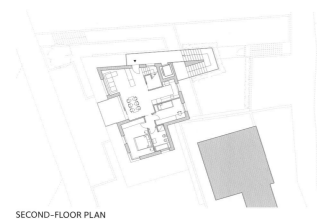
SECOND-FLOOR PLAN

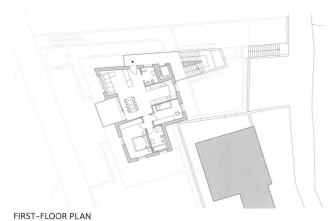
FIRST-FLOOR PLAN

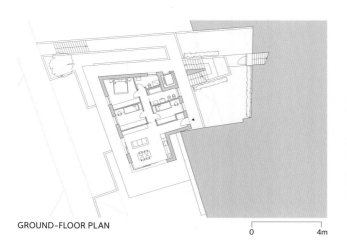
GROUND-FLOOR PLAN

0   4m

The building is classified as A+ (27 kWh/m² per year). Domestic hot water is supplied by a solar-thermal system. To increase energy saving, in addition to the insulation of external walls, ceilings and controlled mechanical ventilation, there has been a focus on rationalizing the internal distribution, providing the building with large openings achieved by using larch low-E windows with triple glazing. These are located at south, east, and west to make the most of the winter sun. The main entrance is situated on the inferior street, while the secondary ones, pedestrian-only, are at the top and bottom of the site. By accessing the building through external spaces the architects have tried to emphasize the use of green elements, obtaining a residential building immersed in the landscape, capable of fostering relationships between the people who live there.

Kjellgren Kaminsky建筑事务所

萨隆根35号

KJELLGREN KAMINSKY ARCHITECTURE
SALONGEN 35

Location Malmö, Sweden **Units** 4 **Area** 7965 ft² (740 m²) **Photography** Kasper Dudzik

Salongen 35, located in the city of Malmö, is the only passive house project in its area. Consisting of four dwellings, the architect used a variation of cladding materials to give each home its own character. Architecturally, the houses are linked by the use of similar detailing, which also helps to give the buildings their human scale and unique expression.

In addition to the project's passive status, the apartments also feature other ecological and sustainable aspects. These include: solar panels; exterior venetian blinds to reduce summer heat and allow winter sun to enter the building; environmentally friendly building materials—all wood is FSC (Forest Stewardship Council) approved; A-rated appliances for minimum energy consumption; and specially designed taps for minimal water usage. Green walls and roofs, flower boxes, and lawns also feature, further enhancing the 'green' focus.

36 SUSTAINABLE HOUSING

GROUND-FLOOR PLAN

FIRST-FLOOR PLAN

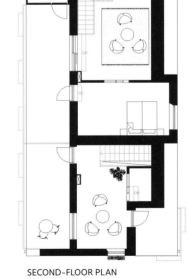
SECOND-FLOOR PLAN

To create distinction among the houses different cladding materials were used—plaster, fiber-cement boards or wood, and aluzinc for the roofs. Flower boxes made from white perforated metal have been placed in selected windows in bedrooms, living rooms, and kitchens. All windows and doors have a thin frame made from different materials for each house.

The interior features a contrasting palette: rusty, earthy, and minimalist. The use of recycled, environmentally friendly materials lowered construction costs. The entrance, laundry, and kitchen floors are made of recycled terra cotta. In the living and dining rooms, the floors are made from rustic-resistant wood cubes of whitewashed pine and the staircase between the floors is in brushed pine. In order to provide surprising views of surrounds and let more light into the apartments, the splayed parts of some windows include mirrors.

Along the west side of the plot is a micro park with plants and benches, creating a varied street space for meeting between residents and visitors.

基兰·汀布莱克
贝尔斯联排房屋

KIERANTIMBERLAKE
BELLES TOWNHOMES

Location San Francisco, California, United States **Collaborating partner** LivingHomes **Units** 7 **Area** 6720 ft² (624 m²)
Photography Richard Barnes/OTTO

Conceived in partnership with Santa Monica-based developer LivingHomes, Belles Townhomes reflect KieranTimberlake's ongoing interest in creating mass-customized, efficiently constructed, and environmentally responsible dwellings. This is the only newly constructed residential building in the Presidio National Park, and the first set of multifamily homes in San Francisco to receive LEED Platinum Certification. Developed as part of an adaptive re-use project, the site is adjacent to former World War I–era officers' quarters and a hospital renovated into 200 rental units. The project demonstrates the evolution of a national historic landmark into a public place defined by 21st-century aesthetic and programmatic functions.

Designed to conserve resources and minimize site impact, the building is confined to a compact 6720-square-foot (624-square-meter) footprint. Seven attached three-story townhomes overlook a shared central green and wooded area. Units are organized to balance private and public zones, remaining visually closed at the base and becoming more open at the upper levels.

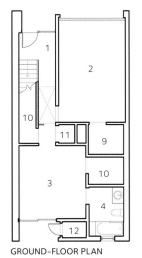 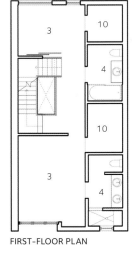 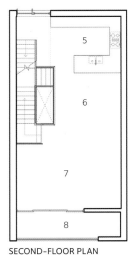 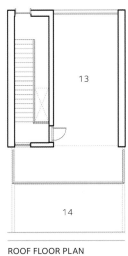

GROUND-FLOOR PLAN　　FIRST-FLOOR PLAN　　SECOND-FLOOR PLAN　　ROOF FLOOR PLAN

1 ENTRY
2 GARAGE
3 BEDROOM
4 BATHROOM
5 KITCHEN
6 DINNING ROOM
7 LIVING ROOM
8 TERRACE
9 MECHANICAL
10 CLOSET
11 LAUNDRY
12 STORAGE
13 ROOF TERRACE
14 GREEN ROOF

BELLES TOWNHOMES

Each three-bedroom, three-bathroom home has outdoor spaces including a patio overlooking the green, a balcony off the main living level, and a panoramic rooftop deck.

Energy use is reduced with efficient appliances and fixtures, radiant heat, and photovoltaic panels. Materials are sustainably harvested and recycled, and include no- or low-VOC carpet, paint, and sealants. Rainwater is stored in an underground aquifer to minimize runoff, and drought-tolerant plants are used in the landscaping. Advanced sensors monitor and manage electricity, water, and gas usage, and provide occupants with real-time feedback on resource use. Over 90 percent of the interior spaces receive natural daylight. The townhomes were designed to achieve LivingHomes' stringent and extraordinary Z6 environmental goals: zero energy, zero water, zero indoor emissions, zero waste, zero carbon, and zero ignorance.

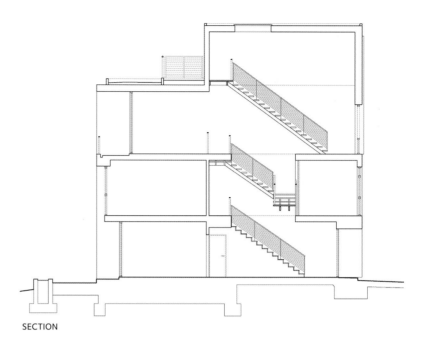
SECTION

贾斯汀·马利亚

扬巷住宅

JUSTIN MALLIA

YAN LANE

Location Richmond, Victoria, Australia **Units** 2 **Area** 3229 ft² (300 m²)
Photography Emma Cross (pages 44, 45, 46, 47 right, 49 top), Paul Cadenhead (page 47 left, 48, 49 bottom)

Undertaken on a small budget, the Yan Lane scheme was conceived as an opportunity to use architectural understanding to drive a development project to meaningfully infill an otherwise ignored space, and to achieve financial return. The project involved the subdivision of a narrow sliver of land with no street frontage, hidden between the rear face of shops to the south, and the backyard fences and sheds of houses to the north. Yan Lane is primarily the creation of a new building, which incorporates two houses that reach beyond the scope of the small site to include the extension of services infrastructure from the main road, and the recreation of a right-of-way to form a new street. The project has turned what was previously a disused and neglected area into an activated, humanized space.

YAN LANE

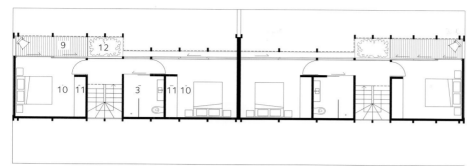
SECOND-FLOOR PLAN

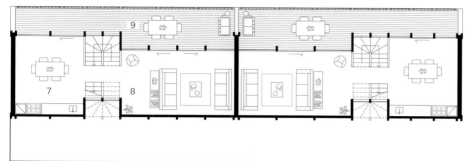
FIRST-FLOOR PLAN

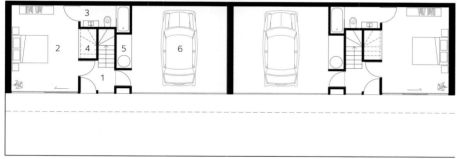
GROUND-FLOOR PLAN

In response to diverse and complex surroundings, the two houses are simply presented as a single building. The form of the building contorts itself in response to the dimensionally tight parameters of the site through a stepped sectional profile. Different materials have been used for each face of the building, and all perform differently to interact with the immediate external context and internal spaces they enclose. A repetitive structural timber frame is exposed and acts as a consistent organizing principle throughout this assemblage, conceptually stitching the façades together to create cohesion.

1. ENTRY
2. OFFICE / STUDIO / THIRD BEDROOM
3. BATHROOM
4. STORAGE
5. LAUNDRY
6. GARAGE
7. KITCHEN / DINING
8. LIVING ROOM
9. TERRACE
10. BEDROOM
11. WARDROBE
12. GARDEN

Toward the noise and clutter at the back of the adjoining shops, the south elevation presents itself in a simple, unified manner through the repetition of expressed timber columns set on a solid zincalume backing. The building can open up to engage with and enliven the laneway or it can close down to a seamless, simple façade. Toward the light, tree canopies and residential character to the north, the envelope is set back from the structural frame. This enables it to be openable with wide sliding doors, becoming a deep, occupiable space through a layering of extensive customized screens, terraces and planting. The façades are flexible, allowing permeability to be mediated to suit the weather or the way the spaces are occupied.

Yan Lane is energy efficient, carefully detailed and tactile—a surprising encounter of light and tranquillity in an otherwise gritty urban setting.

GBD建筑事务所
炉窑公寓

GBD ARCHITECTS
KILN APARTMENTS

Location Portland, Oregon, United States **Units** 19 **Area** 18,000 ft² (1672 m²) **Photography** Eckert & Eckert Photography

The Kiln Apartments project was a research and design effort to develop the most energy efficient apartments possible. Performance goals were ambitious and it was the first new market-rate apartment building in the United States to achieve passive house certification, generating 69 percent fewer emissions than the average US building of the same type and size. Comprising 18,000 square feet (1672 square meters), the apartment block is located in a pedestrian- and bicycle-friendly neighborhood in Portland, Oregon and consists of 19 for-lease apartments and ground-floor retail. Two fundamental goals guided the project team: first, design apartments that retain the charm of Portland's highly desirable inner city homes; and second, achieve dramatic reductions in energy use. The first goal was achieved by designing the apartments to evoke the comfortable qualities of a well-crafted, single-family home, with an eclectic mix of personal touches. Inspired by pleasant childhood memories in the Pacific Northwest, the team used locally sourced materials prioritized for their longevity and aesthetic beauty.

52 SUSTAINABLE HOUSING

Achieving such significant reductions in energy use required strategic design decisions. The primary effort addressed the building's exterior envelope. All apartments are both tightly sealed—meaning exterior air does not infiltrate inside easily—and highly insulated. The building behaves much like a thermos, creating a continuous layer of insulation throughout, keeping the interior temperature relatively stable. The exterior skin seeks to eliminate thermal bridging where heat transfer can reduce energy efficiency. Because of the building's extraordinarily tight envelope, providing mechanically fresh air was critical. There is 24-hour-a-day ventilation through a centralized energy recovery ventilator, delivering high quality indoor air with very low energy use. The Kiln Apartments were honored with an AIA 2030 Challenge Award for Multi-Family Residential Excellence.

KILN APARTMENTS 53

KILN APARTMENTS

ROOF FLOOR PLAN

FOURTH-FLOOR PLAN

FIRST- / SECOND-FLOOR PLAN

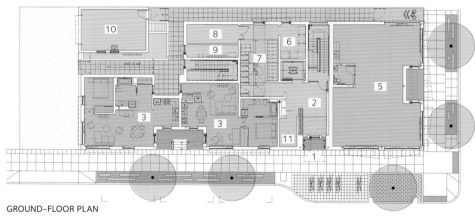

GROUND-FLOOR PLAN

1 ENTRY
2 LOBBY
3 APARTMENT
4 TWO-STORY APARTMENT
5 RETAIL
6 TRASH / RECYCLING
7 BIKE STORE
8 MECHANICAL
9 ELECTRICAL
10 WELDING STUDIO
11 FIRE
12 PUBLIC AMENITY DECK
13 PRIVATE AMENITY DECK
14 SOLAR HOT WATER ARRAY
15 MECHANICAL PENTHOUSE

■ SERVICE
■ HOUSING
■ RETAIL

经济适用公寓

AFFORDABLE APARTMENTS

LAPS建筑事务所与MAB建筑事务所
城市中心公益住房

LAPS ARCHITECTURE + MAB ARQUITECTURA

CIVIC CENTER + SOCIAL HOUSING

Location Paris, France **Units** 30 **Area** 19,741ft² (1834 m²) **Photography** Luc Boegly

This multi-purpose building, located on the corner of Avenue Felix Faure and rue Tisserand, consists of 30 housing units for young workers and a civic center. A symbolic institution in Paris' 15th arrondissement, The civic center was housed until 2010 in offices that became unfit for purpose. In response, the RIVP (Régie Immobilière de la Ville de Paris), a public housing agency, opened a competition to construct a new building that would house the original program but also include 30 studios for young workers. MAB and LAPS won the competition with a holistic conception, highlighting the relationship between the city and public space. The building remains in the same envelope but extends over several different levels, including the two separate identities in a principle of uniqueness. The public facility and the housing enjoy two distinct entrances. The center opens wide onto Avenue Felix Faure, rising one story from the ground floor. Its glassed-in lobby extends the public space inside the building. The residence, on the other hand, has a more toned-down entrance on rue Tisserand, at the side of the building, and rises from the second story to the fifth.

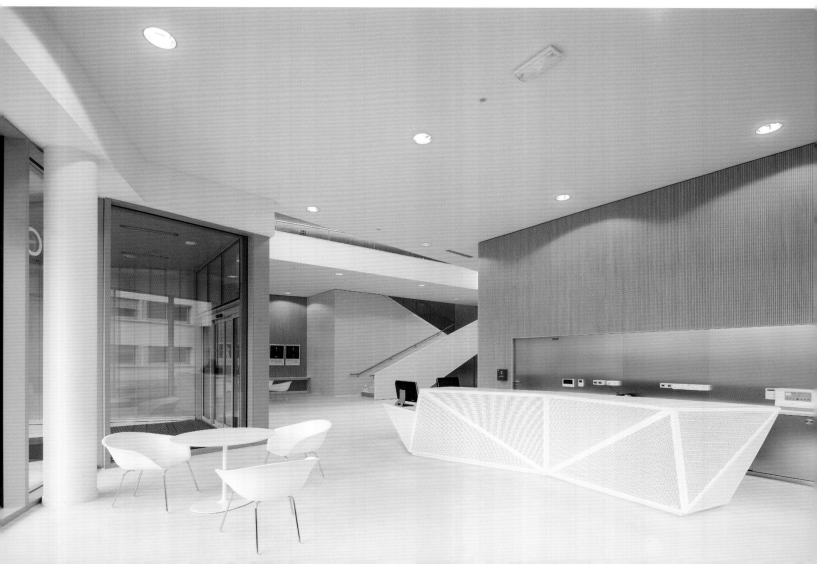

CIVIC CENTER + SOCIAL HOUSING

The four-story façade along rue Tisserand is treated in a light colored, glazed concrete, creating metallic reflections that marry the interior glasswork. The center's reception ash been designed as a covered public space. Its program rises from the ground floor to a technical mezzanine on the first floor, centering around spaces in the double-height lobby. Beyond the computer area and cafeteria, on the ground floor, is a multipurpose hall—a space for entertainment with a stage, control booth, dressing rooms and pocket-sized wings. On the first floor the program includes numerous activity rooms and offices. A patio provides natural light to the art room and the floor's access hallway. Finally, a large room with warm tones offers space for dance, yoga, and low-impact gymnastics.

62 AFFORDABLE APARTMENTS

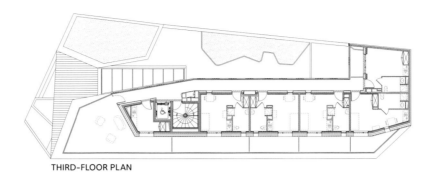

THIRD-FLOOR PLAN

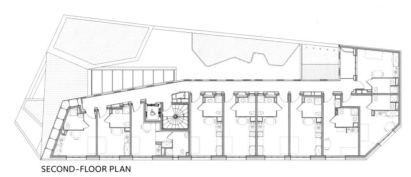

SECOND-FLOOR PLAN

FIRST-FLOOR PLAN

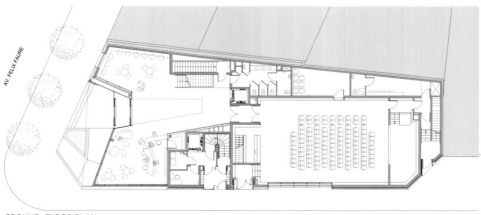

GROUND-FLOOR PLAN

RUE TISSERAND

0   5ft

The living units, varying from 205 to 248 square feet (19 to 23 square meters) combine a working wall (storage, kitchen, office) and naturally lit, ventilated bathrooms. Generous windows dilate the studio space and a community terrace on the building's fourth floor overlooks Paris, dispelling preconceived ideas about joint ownership in social housing. Circumventing the prospect rules made it possible to offer each unit a private terrace.

With the Eiffel Tower visible in the distance this radical and ambitious building serenely proclaims that state-funded buildings in the 21st-century can indeed be lived in.

rh+建筑事务所

普莱因·苏蕾公寓

RH+ ARCHITECTURE

PLEIN SOLEIL

Location Paris, France **Units** 28 **Area** 27,555 ft² (2560m²) **Photography** Luc Boegly

Situated on an exceptional block widely visible from the corner of Avenue de Flandre, Plein Soleil is very close to the Bassin de la Villette, Paris' largest lake, and benefits from 118 feet (36 meters) of south-facing frontage.

In order to take advantage of the southern front line, a thermal façade with loggias was designed to regulate temperature for comfort during summer and winter months. This external buffer space consists of two sliding-glass windows, which allow heat from sunlight to be absorbed by the floor and walls during the day and released at night. In colder weather, fresh air provided by controlled mechanical ventilation is preheated.

The multi-oriented apartments have been designed with entrance and bathroom to the north, and rooms with balconies to the south, with some flats containing double-height living rooms.

The west building has been developed in a stepped shape and provides the private terraces and green roofs with a west-facing view, creating favorable external spaces for all.

PLEIN SOLEIL 67

68 AFFORDABLE APARTMENTS

GROUND-FLOOR PLAN

FIRST-FLOOR PLAN

FIFTH-FLOOR PLAN

1 CHILDCARE CENTER
2 ENTRANCE
3 STUDIO
4 TWO-ROOM APARTMENT
5 THREE-ROOM APARTMENT
6 FOUR-ROOM APARTMENT
7 SERVICE ROOM
8 CAR PARK
9 COURTYARD
10 TERRACE

大卫·贝克建筑事务所
理查森公寓

David Baker Architects
RICHARDSON APARTMENTS

Location San Francisco, California, United States **Units** 120 **Area** 65,419 ft² (6078 m²)
Photography Bruce Damonte (pages 70, 72, 73, 74, 75 top), Mathew Millman (pages 71, 75 bottom)

Richardson Apartments provides supportive housing for formerly homeless individuals living in San Francisco. This 120-unit building provides various on-site services to help residents overcome social, medical, and employment challenges as they transition to new phases of their life. The project also contributes to the Market and Octavia Neighborhood Plan, which aims to create a dense transit-oriented district with retail corridors and pedestrian-friendly streets that will boost the local economy and revitalize a formerly diverse and thriving community.

The project was driven by the architects' belief that everyone deserves to live in a well-designed space. The design pays specific attention to creating a building that is not only functional and aligned with residents' needs, but also helps foster an atmosphere of transformation; a place in which residents will feel safe, stable, and supported.

RICHARDSON APARTMENTS

75

GROUND-FLOOR PLAN

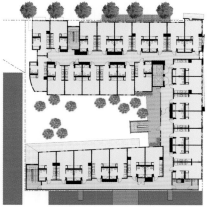
FIRST-FLOOR PLAN

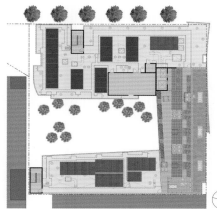
ROOF FLOOR PLAN

The architects focused on designing a building that embodied beautiful modern architecture in order to empower residents to have pride in their building and see both it and themselves as valuable additions to the neighborhood.

The project design has been leveraged as a tool to transform lives, build new communities, and respond to community context. The building is sited in a central transit-rich neighborhood and forgoes auto parking, instead maximizing the space and construction budget for more units and increased common spaces. At the urban scale, Richardson Apartments will increase local property value and reduce the total number of people experiencing homelessness in the city.

凯文·戴利建筑事务所
百老汇住房

KEVIN DALY ARCHITECTS
BROADWAY HOUSING

Location Santa Monica, California, United States **Units** 33 **Area** 33,225 ft² (3086 m²) **Photography** Nico Marques

Developed by the Community Corporation of Santa Monica (CCSM), Broadway Housing addresses the housing needs of families earning from 30 to 60 percent of the local median income. The developer is committed to building and operating infill housing that is environmentally and economically sustainable in a city known for progressive politics and daunting development regulations.

The project replaces a vacant nursing home, adding density and activity to an urban corner site across from a large community park. The design aggregates two- and three- bedroom units into four repeatable blocks arranged in a pinwheel configuration around the site. After-school programs are offered in a cluster of community buildings with planted roofs.

80 AFFORDABLE APARTMENTS

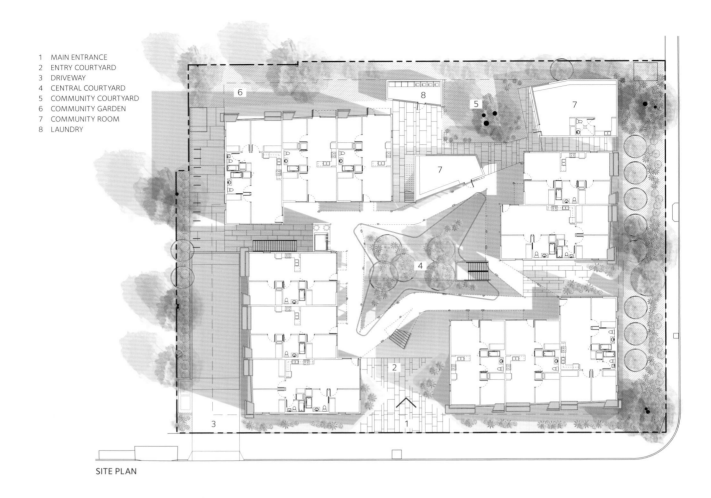

1. MAIN ENTRANCE
2. ENTRY COURTYARD
3. DRIVEWAY
4. CENTRAL COURTYARD
5. COMMUNITY COURTYARD
6. COMMUNITY GARDEN
7. COMMUNITY ROOM
8. LAUNDRY

SITE PLAN

The three-story residential buildings are scattered around a starfish-shaped internal courtyard with a planter that extends through the underground parking level, allowing mature sycamore trees to thrive and shade the courtyard. The four residential blocks are connected by a multilevel, adding layers of privacy yet providing views across the courtyard. All of the units open to the central courtyard, providing natural ventilation through each dwelling.

Sustainable design strategies are central to the goal of building durable housing that is economical to operate. The project is naturally ventilated, so controlling daytime heat was a top design priority. This is achieved through planted roofs, which insulate community buildings and minimize heat accumulation in spaces used for late afternoon programs. The sun-facing 'active' façade of each building is composed of custom window boxes, which protect glazing from direct sun. The entire site is engineered to collect all roof and surface rainwater, piping it to an underground 15,000-gallon (57-cubic-meter) cistern where it is clarified and used for landscape irrigation. The project has been recognized with a national AIA Honor Award, an AIA/CC Design Award, and an AIA/LA Design Award.

乔纳森·西格尔，美国建筑师协会会员
北帕克公寓

JONATHAN SEGAL FAIA
THE NORTH PARKER

Location San Diego, California, United States **Units** 25 **Area** 950 ft² (88 m²) **Photography** Matthew Segal

Designed, developed, and owned by Jonathan Segal FAIA, The North Parker project proves there is still opportunity to construct affordable accommodation among market-rate housing, while activating an up-and-coming neighborhood. The North Parker project is now the southern gateway to the newest transitioning neighborhood in San Diego. The corner of 30th and Upas Streets, previously blighted by barely standing structures and a propensity towards vagrants, is now a community gathering point. The mixed-use project consists of 25 market-rate units, two low income units, five commercial spaces; which consist of three restaurants, a beer-tasting bar, and an architectural office. Simple forms and a simple palette create a powerful presence on the street corner. The commercial spaces open up the ground plane to separate the residences from the busy street activity.

84 AFFORDABLE APARTMENTS

THE NORTH PARKER 85

Multiple entrances through different nodes of the project allow people to transfer between the commercial ground plane along the street, to the interior garden, courtyard space, and then up stairs to the second-level exterior circulation paths. Tenants are able to enter their units through semi-private exterior patios, which define their space. These patios have been raised two-feet above the walkway to create a feeling of privacy while maintaining visual connection. A constant relationship between street activity and neighboring facilities means both commercial and residential are interlaced together.

AFFORDABLE APARTMENTS

THE NORTH PARKER

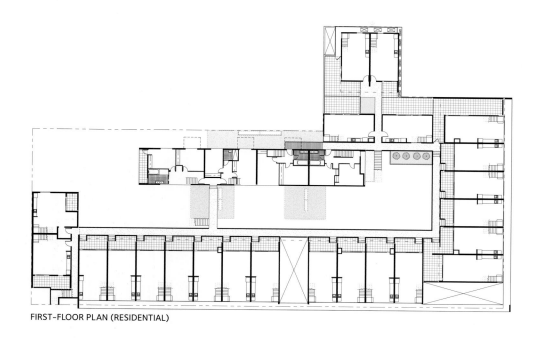

FIRST-FLOOR PLAN (RESIDENTIAL)

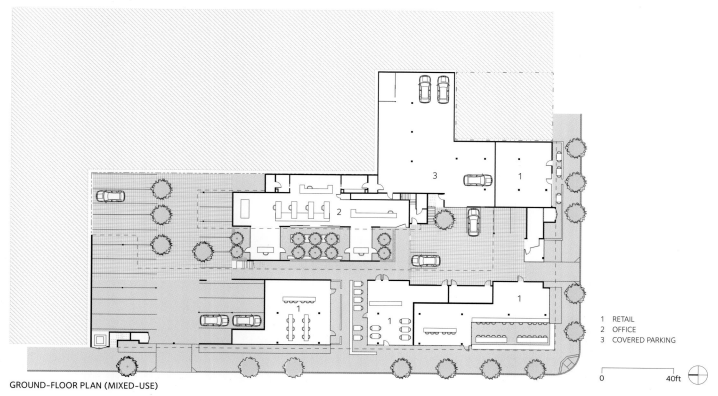

GROUND-FLOOR PLAN (MIXED-USE)

1 RETAIL
2 OFFICE
3 COVERED PARKING

吉耶尔莫·巴斯克斯·孔苏埃格拉建筑事务所
瓦尼卡斯公益住房

GUILLERMO VÁZQUEZ CONSUEGRA ARCHITECT

SOCIAL HOUSING IN VALLECAS

Location Madrid, Spain **Units** 165 **Area** 275,039 ft² (25,552 m²) **Photography** Duccio Malagamba

The Social Housing project, located in Vallecas, Madrid, proposed the construction of two parallel blocks instead of the enclosed precinct recommended by the expansion area planning regulations.

Uneven conditions in the surrounding area as well as the allotment's urban context and aspect made the parallel layout advisable. The aim was to construct better homes with improved orientation, to allow for nicer views.

The two equally parallel blocks are aligned with the outer ends of the allotment, leaving a large garden area in the middle. In this arrangement, some of the homes face outwards and others overlook the garden. All have two outer façades ensuring natural cross-ventilation and lighting for every part of the dwelling.

AFFORDABLE APARTMENTS

SOCIAL HOUSING IN VALLECAS

Diverse exterior situations outside the homes generate a variety of 'skins'. The smooth, continuous walls are pierced by long windows on the outer wall. Corridors running along the inner wall are protected by aluminum uprights and provide views of the garden, framing a dynamic and changing landscape.

Parking and storage spaces are set directly below ground level, beneath the apartment blocks. Despite enormous difficulties involved with sharing the same structural framework as the apartments, it was necessary in order to free up the space between the blocks and shape a real garden.

The two blocks are by no means identical. One of them has an angled cut at one end that adapts to the allotment geometry facing the traffic roundabout. At the other end, a subtraction operation on the volume marks the entrance to the complex.

The apartment distribution strives to ensure versatile and qualified interior spaces, which comply with the functional organization requirements set out in the brief and planning regulations, as well as permitting greater spatial and structural flexibility for alternative uses of the homes.

SOCIAL HOUSING IN VALLECAS

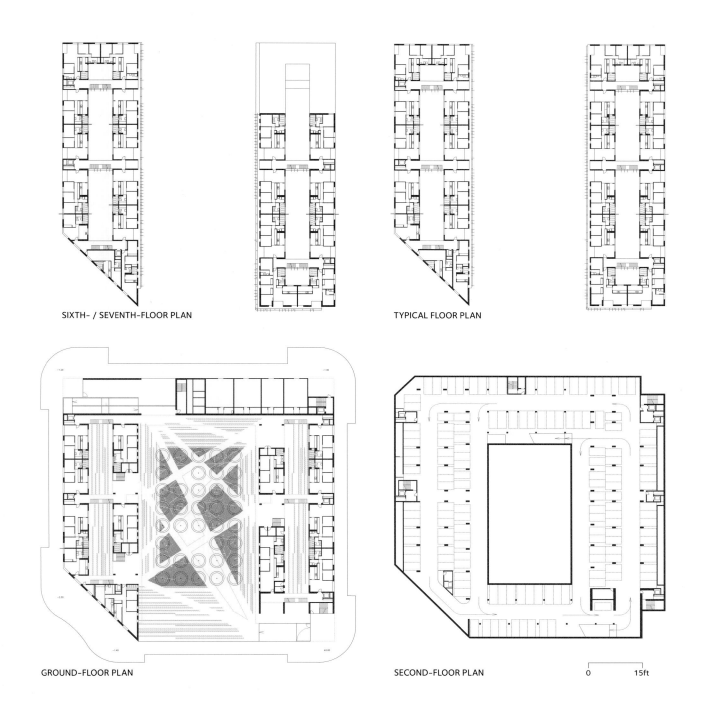

SIXTH- / SEVENTH-FLOOR PLAN

TYPICAL FLOOR PLAN

GROUND-FLOOR PLAN

SECOND-FLOOR PLAN

蒂普尔建筑事务所
里士满街60号

TEEPLE ARCHITECTS
60 RICHMOND STREET

Location Toronto, Canada **Units** 85 **Area** 99,565 ft² (9250 m²)
Photography Shai Gill (pages 94, 96–8, 99 top), Scott Norsworthy (pages 95, 99 bottom)

This project is an 11-story, 85-unit housing cooperative and mixed-use building located in Toronto's downtown precinct. The client program—to provide affordable housing for hospitality workers—was a key inspiration for the design, which incorporates social spaces dedicated to food and its production. The result is a small-scale, full-cycle system described as 'urban permaculture,' where the resident-owned and operated restaurant and training kitchen on the ground floor is supplied with vegetables, fruit, and herbs grown on the sixth-floor terrace. The kitchen garden is irrigated by stormwater, which is collected on the roofs, while organic waste generated by the kitchens serves as compost for the garden.

60 RICHMOND STREET 97

AFFORDABLE APARTMENTS

The project was conceived as a solid mass where volumes were subtracted and carved away in order to create openings, a courtyard space, and terrace at various levels. This method of deconstructing the volume creates spaces that step out and back from the street, interlocking and contrasting with each other. It was also crucial in achieving several key objectives: the creation of the sixth-floor garden, drawing light into the interior spaces, and providing outdoor green space. The client's request for low maintenance costs inspired many sustainable initiatives, including cladding the exterior of the building in fiber-cement panels, high-performance windows, a sophisticated mechanical system, and heat recovery. A low-maintenance green roof and rainwater collection system also help to reduce the carbon footprint.

Seeking to imagine the city as an extension of the natural environment, 60 Richmond Street explores the notion of urban form as environmental form. Teeple Architects provide an example for future urbanism, with imaginative architectural solutions effectively addressing global environmental challenges. This project is an iconic design showcasing a radical and innovative approach to urban infill.

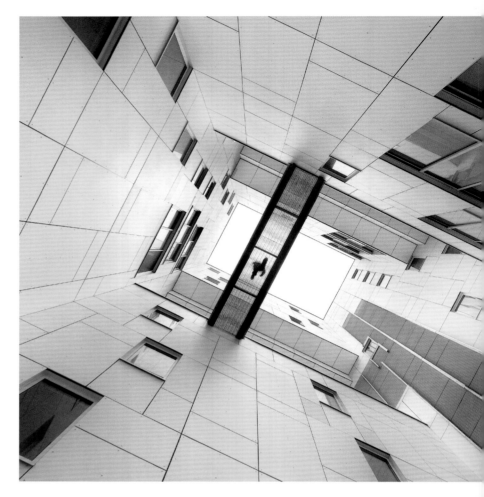

60 RICHMOND STREET

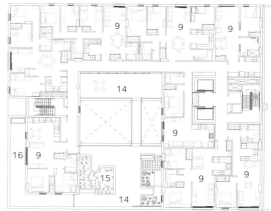

SIXTH–EIGHTH-FLOOR PLAN

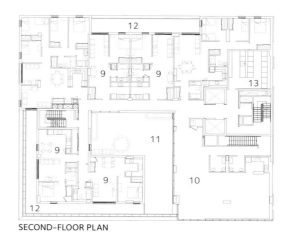

SECOND-FLOOR PLAN

GROUND-FLOOR PLAN

0 10m

1 RESIDENTIAL LOBBY
2 CO-OP OFFICES
3 RESTAURANT
4 KITCHEN
5 PANTRY
6 BICYCLE PARKING
7 PARKING AREA
8 GARBAGE/SERVICE SPACE
9 RESIDENTIAL UNIT
10 INDOOR AMENITY SPACE
11 OUTDOOR AMENITY SPACE
12 PRIVATE TERRACES
13 LAUNDRY ROOM
14 PUBLIC TERRACES
15 GARDENS
16 PRIVATE TERRACES

卡萨诺瓦事务所与赫尔南德斯建筑事务所
黑白双子公寓

CASANOVA+HERNANDEZ ARCHITECTS
BLACK & WHITE TWINS

Location Blaricum, Netherlands **Units** 29 **Area** 26,480 ft² (2460 m²) **Photography** Christian Richters

In Blaricummermeent, an exclusive urban development in Blaricum, the Netherlands, the most common form of architecture is single-family villas and houses. It is within this context that the design of Black & White Twins experiments with spatial and visual integration of compact and selective housing solutions. Designed for Blauwhoed Eurowoningen, each of the twin volumes are four levels high and comprise 14 and 15 apartments, respectively, within a unique visual integration of suburban context.

Each of the volumes are conceptually divided into two sections: an outer black skin—creatively perforated with windows and voids—that wraps around the inner part of the building, and the open-air spaces that are colored in white. This chromatic contract emphasizes the independence of the skin layer of the building, which is responsible for contextualizing and integrating the visual aesthetic of the building within its urban surroundings.

AFFORDABLE APARTMENTS

BLACK & WHITE TWINS 105

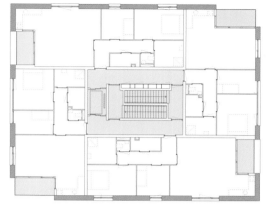
BLOCK B THIRD-FLOOR PLAN

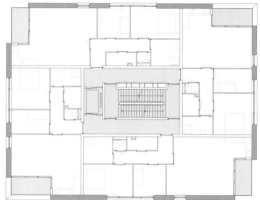
BLOCK B SECOND-FLOOR PLAN

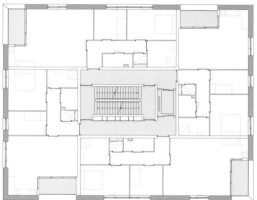
BLOCK B FIRST-FLOOR PLAN

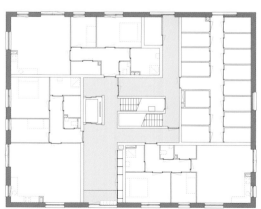
BLOCK B GROUND-FLOOR PLAN

1 BLOCK A
2 BLOCK B

SITE PLAN

From inside, this outer layer differentiates each of the apartments by providing a unique interface—intentionally placed windows of alternating sizes—that shapes the view to the neighborhood outside. This playful rhythm has been reproduced internally, by alternating the floor plans of the apartments. Each is unique, and although the apartment layouts share commonalities, each is presented with a differing size and organization of living areas in order to promote the inherent diversity of the project. These twin buildings provide affordable apartments within small-scale housing blocks, minimizing their intervening footprint and physical impact on the residential area while encompassing a highly unique visual space.

豪华住宅

LIVING IN
LUXURY

Aytac建筑事务所

18号公寓

AYTAC ARCHITECTS
APARTMAN 18

Location Istanbul, Turkey **Units** 9 **Area** 29,062 ft² (2700 m²) **Photography** Cemal Emden

Apartman 18 is a 10-story residential building containing nine non-identical apartment units, rooftop garden with swimming pool, street-level garden, underground parking, and shared spaces. It is in the residential neighborhood of Erenköy, in Istanbul's Asian quarters. Apartman 18 pays homage to Erenköy's vineyards, which were completely destroyed due to densification of the city with concrete cookie-cutter apartment blocks after the 1970s.

The volume comprises one continuous surface around which the 10 varying floor plates are spirally wrapped. This surface has a three-dimensional façade consisting of individually shaped aluminum panels offering privacy with maximum natural light and external views. The 'vine-like' façade moves upward to become the intertwining building façade. It begins at a roof garden on the uppermost level and emerges at the ground level, functioning as a landscape element and creating meditative gardens at both top and bottom. Within the neighboring buildings, the dance of intertwining surfaces creates a pleasant surprise.

LIVING IN LUXURY

Internally, the hallway that connects the living and sleeping quarters exhibits varied degrees of translucency to create a visual connection with the city without compromising privacy. It also creates a cinematic effect by animating the façade. High-efficiency air-to-water heat pumps are used for heating and cooling the apartments and hot water. Apartman 18 is the first of its kind in Turkey to utilize external air as a renewable energy source, avoiding the use of any fossil fuels, which dramatically reduces the energy costs and minimizes the volume's impact on the environment. These efforts are further propelled by the use of insulated glass units with double silver coatings, which ensure further energy savings and a reduction in solar heat gain. The combined effect of the insulated glass and aluminum panels of the façade also allows for protection from electro-magnetic smog prevalent in this area.

LIVING IN LUXURY

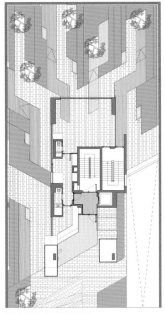

GROUND-FLOOR PLAN

FIRST-FLOOR PLAN

SECOND-FLOOR PLAN

THIRD-FLOOR PLAN

FOURTH-FLOOR PLAN

FIFTH-FLOOR PLAN

SIXTH-FLOOR PLAN SEVENTH-FLOOR PLAN EIGHTH-FLOOR PLAN NINTH-FLOOR PLAN ROOF FLOOR PLAN 0 5m

里伯斯金工作室
维特拉公寓

STUDIO LIBESKIND
VITRA

Location São Paulo, Brazil **Units** 14 **Area** 159,844 ft² (14,850 m²)
Photography Ana Mello (pages 114, 115, 118, 119), Romulo Fialdini (pages 116, 117)

Vitra is a luminous glass-clad, high-rise residential project in the Itaim Bibi district of São Paulo. Located near a number of the city's main thoroughfares the project offers access to the popular Ibirapuera and do Povo parks. The bold sculptural design features a multi-faceted glass façade articulated by inlaid balconies, which form a rhythmic pattern across the façade.

According to the architect, 'The inspiration for the project is the city of São Paulo and the Brazilian people. I designed this tower to expresses the optimism, vibrant culture, and dynamic possibilities of a truly pluralist society.'

Vitra comprises 14 floor-through residences and a penthouse, each featuring a unique floor plan ranging in size from 6080 square feet (565 square meters) to 12,325 square feet (1145 square meters). The sun-soaked lobby features a cast-in-place concrete reception desk, which is set against a floor-to-ceiling Brazilian wood wall. The pairing of concrete and warm wood is echoed throughout the building's clean and modern interiors.

VITRA 117

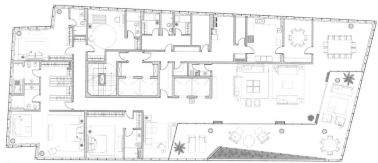
SECOND-FLOOR PLAN

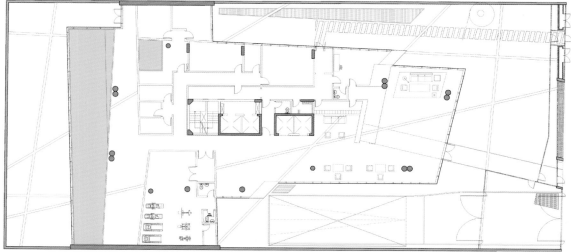
GROUND-FLOOR PLAN

Interior designer Dado Castello Branco outfitted the interiors with artworks such as *Ultramar Intervalo* by famed Rio de Janeiro artist José Bechara. The building's amenities include a swimming pool, spa, gym, multipurpose lounge, and a playroom for children.

Since its inception, one of the pillars of the project has been the adoption of sustainable practices. The environmental solutions employed in Vitra include a system for rainwater collection and reuse; solar panels for heating water; the use of sustainable materials, which reduce the energy consumption caused by air conditioning and elevators; energy-efficient, low emissivity glass, which impedes thermal transmission; intelligent building management systems; and the efficient management of rubble generated during construction.

'Vitra is a new concept in condominium residences. With its sculptured, crystalline form, it creates a new icon for the city of Sao Paulo', affirms Paulo Oliveira, president of JHSF's incorporation business unit (a Brazil-based developer in high-end real estate).

彼得·毕希勒建筑事务所
镜屋

PETER PICHLER ARCHITECTURE
THE MIRROR HOUSES

Location South Tyrol, Italy **Units** 2 **Area** 861 ft² (80 m²) **Photography** Oskar Da Riz

Set in the marvellous surroundings of the South Tyrolean Dolomites, amidst the apple trees, are The Mirror Houses. Combining the highest standards of contemporary architecture with one of the most astonishing landscapes nature has to offer, these two luxury apartments provide guests with a truly unique vacation experience.

The client, who lives on site in a refurbished farmhouse from the 60s, wanted a structure that could be rented out as luxury holiday units with a focus on achieving maximum privacy for both resident and guest. The building comprises two units, each containing a kitchen/living room, a bath and bedroom with large skylights that open to allow natural light and ventilation, and a small basement for temporary storage. Because the initial volume of the structure has been split in two, the units slightly shift in height and length in order to loosen the entire structure and articulate each space. Both units sit on a floating base aboveground to evoke a sense of lightness and offer better views from cantilevering terraces.

124 LIVING IN LUXURY

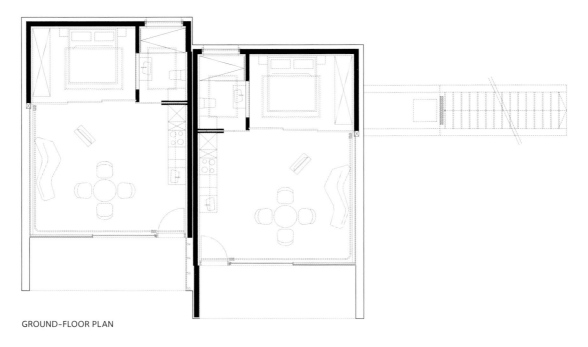

GROUND-FLOOR PLAN

The volume opens toward the east with a large glass façade that fades with curvilinear lines into the black aluminum shell. Mirrored glass on the west façade of the units borders the client's garden and catches surrounding panoramic views, while making the units almost invisible. The mirrored glass is laminated with an ultraviolet coating, which prevents bird collision. Certain views from the residential garden see the existing farmhouse mirrored and literally blending into the new contemporary architecture, rather than competing against it.

本杰明·哈尔设计公司
白石工作室

BENJAMIN HALL DESIGN
WHITE STONE STUDIOS

Location Phoenix, Arizona, United States **Units** 6 **Area** 2700 ft² (251 m²)
Photography Lindsey Averill (pages 126, 128 top, 129, 130, 131 top), Winquist Photography (pages 127, 128 bottom, 131 bottom)

White Stone Studios is a small multifamily property designed to provide an alternative rental housing option for downtown Phoenix residents. The luxury micro-rentals were designed and built for young professionals not yet able to afford a dream home but who value quality over quantity within the spaces they inhabit. Comprising a small owner/architect and contractor team, the project engages the neighborhood, site, and building materials in a thoughtful and diligent manner, maximizing both usage and comfort.

The project has been praised for its use of highly accessible, supply-chain, pre-manufactured products, which are often hacked into semi or completely custom elements. Because it's a rental property, affordable and replaceable materials and furnishings were selected, while still building elegant solutions. Materials used in the studio spaces often perform a double-duty. The kitchen island countertops function as both a dining area and work surface, with all setbacks utilized in communal and private outdoor spaces.

130　LIVING IN LUXURY

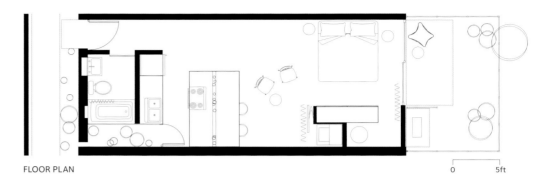

FLOOR PLAN　　　0　5ft

WHITE STONE STUDIOS 131

Nearly all of the building's exterior materials are locally sourced, from the custom masonry blocks to the reclaimed-pallet wood fencing, and the desert vegetation. The masonry blocks form an all-in-one exterior and interior wall, which eliminates the cost of paint and drywall. The blocks also contain a white admixture that helps reflect heat off of the outside of the building while brightening the inside.

On receiving the 2015 AIA Arizona Distinguished Building Honor Award a member of the 2015 selection jury described the project as 'a transformation of ordinary everyday architecture into an extraordinary value proposition. One that combines artisanal attention to craft and detail with strategic thrift, and brings quality design within reach of younger clientele.' White Stone Studios has become an anchor for the neighborhood, influencing a tremendous wave of positive growth in the area.

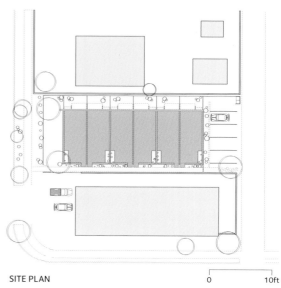

SITE PLAN   0   10ft

比罗^普罗伯茨建筑事务所
锡尔特公寓

BUREAU^PROBERTS
SILT APARTMENTS

Location Brisbane, Australia **Units** 7 **Photography** Christopher Frederick Jones

Situated beneath Brisbane's iconic Story Bridge in Kangaroo Point, SILT is an eight-story luxury apartment block. The architect wanted to create an apartment building that offered something different to Brisbane's high-end residential market. The result is regarded as being highly responsive to its location, connecting with the Brisbane River and complementing the historic Story Bridge and surrounding landscape in scale, form, and materiality. The whole-floor apartments are orientated to maximize views, while openings, screens, and hoods reduce sun exposure and noise from traffic. The west, south, and north aspects of the rectangular form are dominated by precast concrete massing, which offers optimal protection from traffic pollution on Story Bridge.

TYPICAL FLOOR PLAN

1 ENTRY
2 LIFT
3 STAIRS
4 BALCONY
5 KITCHEN
6 DINING
7 LIVING
8 LAUNDRY
9 BEDROOM 1
10 WALK-IN-ROBE
11 EN SUITE
12 STUDY
13 BATHROOM
14 BEDROOM 2
15 BEDROOM 3

Consisting of three bedrooms and study, the apartments have been designed around a central pod, which houses the kitchen, laundry, and powder room. The bedrooms and living area are also designed around this pod. The internal space is extended by a large balcony, which can be completely covered or uncovered by sliding louvre screens. All the internal finishes are based on a natural palette in both color and material. Externally, the linear, black-tinted windows and precast concrete façade are a departure from typical multi-residential structures in the area. SILT was awarded Queensland's highest accolade for Residential Architecture—Multiple Housing, by the Australian Institute of Architects, and was shortlisted in the prestigious World Architecture Festival (WAF) Awards.

艾伦伯格·弗雷泽
318号住所

ELENBERG FRASER
ABODE318

Location Melbourne, Australia **Units** 450 **Area** 484,375 ft² (45,000 m²) **Photography** Peter Clarke Photography

In such a unique location—the bustling cultural center of Melbourne—the design of Abode318 aims to give each occupant maximum views and allow each apartment to be brightened with full natural light. Along the 55-story, wave-like structure, each of the horizontal and vertical 'waves' consist of individual rooms articulated as protrusions. These punctuations along the façade allow each apartment a corner view along the promenade of Russell Street, challenging conceptions of the homogeneity and limitations of apartment city living.

Abode318 comprises a varied range of apartment designs, with over 80 unique floor plans in one-, two- and three-bedroom configurations. The ninth and 55th floors are dedicated to shared amenities—a black granite infinity pool, sauna, steam room, terrace, meeting rooms, dining and lounge areas, and a fully-equipped gymnasium, which overlooks the stretching city of Melbourne.

ABODE318   143

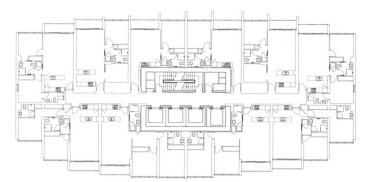

TYPICAL FLOOR PLAN

The volume achieves a unique softness that is not often achieved with skyscrapers that rely on vertical façades. The curve of Abode318 is an innovation in construction technology that not only allows individual expression of each apartment but also contributes aesthetically to the CBD of Melbourne. Additionally, the curved exterior is clad with a low-emissivity glass, which invites maximum natural light saturation while appearing with a pink blush from the exterior, giving the volume a distinctive appearance. The base of the building incorporates industrial design through the detailing of the decorative mesh screen surrounding it.

Inside, the interiors exhibit a sophisticated luxury, with a palette of silver and ivory communal areas, travertine walls, and apartments designed by Australian creative design agency Disegno. This apartment building offers highly stylized comfort and amenity within the bustling cultural hub of Melbourne's CBD.

MA建筑事务所与Neometro建筑事务所
威尔士街公寓

MAARCHITECTS + NEOMETRO
WALSH STREET APARTMENTS

Location South Yarra, Victoria, Australia **Units** 4 **Photography** Derek Swalwell

Walsh Street continues a long tradition of collaboration between Neometro and MAArchitects. Positioned in a street home to houses by Robin Boyd, Frederick Romberg, and Neil Clerehan, The Walsh Street apartments continue the same modernist themes of simplicity and refined architectural expression.

The brief was for four units that would appeal to downsizers and families, blurring the lines between a house and an apartment.

Each apartment occupies a single level of the building and has been designed on similar principles to a courtyard house. The internal planning is orientated around a series of elevated courtyard balconies, which afford external aspects but maintain privacy.

WALSH STREET APARTMENTS 149

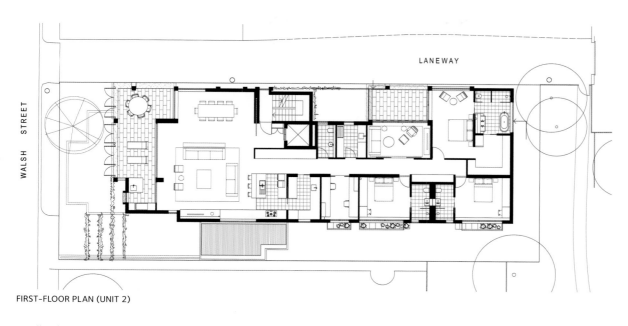

FIRST-FLOOR PLAN (UNIT 2)

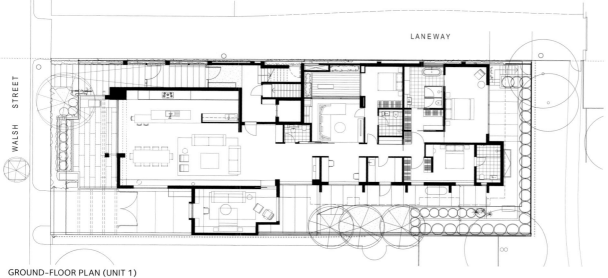

GROUND-FLOOR PLAN (UNIT 1)

The generous interior volumes emphasize a constant connection with outdoor space, while considered architectural details respond to a myriad of overlooking requirements in subtle and effective ways. The result is four unique and uncompromised residences on a sensitive in-fill site.

Interiors by Carr Design reflect seamless joinery and a simple material palette, providing the perfect environment for residents to make their own.

External materials include a combination of concrete, natural renders, and stone cladding. Louvre shutters onto Walsh Street provide an active street façade and occupant control of both sun and privacy, allowing year-round use.

创意社区

CREATING COMMUNITIES

斯泰普尔顿·艾略特设计公司

马歇尔巷公寓

DESIGNGROUP STAPLETON ELLIOTT

MARSHALL COURT

Location Wellington, New Zealand **Units** 27 **Area** 5888 ft² (547 m²) **Photography** Paul McCredie

Marshall Court is a four-story, T-shaped formed block located on a corner site at the edge of Miramar's suburban center in Wellington, New Zealand. Comprising of 27 one-bedroom apartments, it has been designed especially to house superannuitants (65+ years). Placement allows the building to act as a visual 'bookend', mediating between the commercial zone and surrounding residential area. With a well-serviced bus route and suburban conveniences all within walking distance, it is convenient for older or less mobile residents.

The brief required the exisiting and outdated building to be demolished and replaced with a new design, which fitted within the bulk and location of the old building's envelope. The 484-square-foot (45-square-meter) units were designed to feel larger, with sliding screens between the bedroom and living areas allowing for flexible spatial organisation. Apartments are organized into living, sleeping, and service zones, with bedrooms orientated to the street and the kitchen, and service and entry orientated to the breezeway.

MARSHALL COURT 155

MARSHALL COURT 157

FLEXIBLE LIVING ZONES

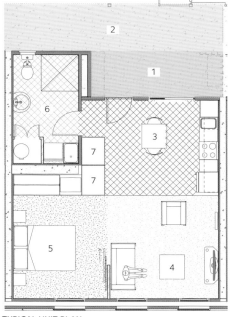

1 VERANDAH
2 BREEZEWAY
3 KITCHEN / DINING
4 LIVING
5 BEDROOM
6 BATHROOM / LAUNDRY / SERVICES
7 STORAGE

TYPICAL UNIT PLAN

The redeveloped complex received regional and national awards (multi-unit dwelling category) from the New Zealand Institute of Architects in 2015. This highlighted the smart use of space, with circulation radiating efficiently to outdoor terraces at the ends of the balconies and north-facing alcoves at each entranceway. These spaces mediate between the private space and public breezeways, providing opportunities for residents to personalize their space and to enjoy social interaction. The breezeway circulation routes and stairs overlook north-facing landscaped gardens.

The new building is designed to fit well into the Miramar neighborhood and provides its residents with a safe, secure, and modern living environment as well as purpose-built community facilities including a social room, BBQ area, and allotment gardens.

1 ENTRANCE
2 COMMUNAL ROOM
3 CENTRAL LIFT
4 APARTMENT
5 CAR PARK
6 MODULATED WALL
7 BREEZEWAY

GROUND-FLOOR PLAN

TYPICAL FLOOR PLAN

YUUA建筑设计事务所

NOIE集合住宅

YUUA ARCHITECTS & ASSOCIATES

NOIE COOPERATIVE HOUSE

Location Tokyo, Japan **Units** 11 **Area** 10,537 ft² (979 m²) **Photography** Toshihiro Sobajima

Due to the density of location, NOIE Cooperative House required a compact design solution. Situated in the middle of an urban block surrounded by small houses, it is only accessible via a 66-foot-long (20-meter-long) corridor. The overall structure is composed of town-house style buildings, which are all connected and resemble the *machiya* (traditional Japanese townhouse) style. Through this design, the unique composition of each house appears directly in the façade and shapes the townscape.

NOIE owes its uniqueness to a flexible design strategy, whereby the building process is as follows: the developer chooses a site, the architect devises a skeleton plan, then future residents take part in designing personalized dwellings with the design team. This infill design was developed and based on a structural framework devised in the initial plan. The aim was to come up with a framework that could provide various functions. The challenge was to develop the space dynamically through interactivity with the skeleton frame and infill design, instead of simply adjusting the infill design into the skeleton frame.

160 CREATING COMMUNITIES

In order to achieve a space unique to the lifestyles of residents, the following code was proposed to allow for design freedom and flexibility of the skeleton frame: a town-house typology unit layout, providing independency for each household and free-choice of floor level; 129 square feet (12 square meters) of utility space, which residents are at liberty to determine the usage of; and rooftop terrace, with variations in floor level to create a dynamic environment.

NOIE COOPERATIVE HOUSE 161

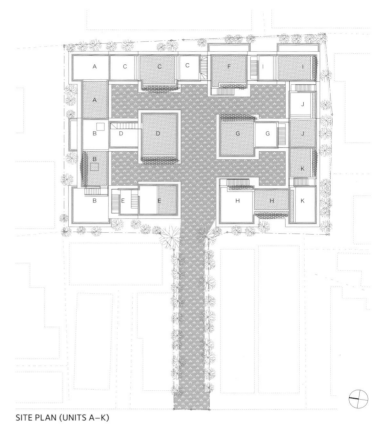

SITE PLAN (UNITS A–K)

NOIE COOPERATIVE HOUSE

麦卡曼特&达雷特建筑事务所
狼溪住宅

MCCAMANT & DURRETT ARCHITECTS
WOLF CREEK LODGE

Location Grass Valley, California, United States **Units** 30 **Area** 3500 ft² (325 m²)
Photography Courtesy of the architect (pages 164, 165, 166 bottom, 168, 169), Ed Asmus (pages 166 top, 167)

A major design goal for Wolf Creek Lodge was to optimize community environmental and social sustainability. Presenting many unique opportunities, the hillside site provided a transition from the adjacent commercial areas and orientates residences toward views of the surrounding woods, while defining an important neighborhood street corner. Its southern orientation also embraces the sun, encouraging use of solar panels and passive heating-and-cooling techniques. Balconies run along the entire south façade, central terrace, and garden court, providing resident privacy and the opportunity for social interaction.

The lodge includes 30 one- and two-bedroom independent condominiums and 3500 square feet (325 square meters) of common facilities. These include a large dining room, gourmet kitchen, sitting area with fireplace, laundry, office, crafts rooms, and three guests rooms, one of which can be used as caregiver quarters. Outside there are community gardens and a spa, as well as 35 parking spaces in a variety of open and garage areas. The community itself is located on less than three acres (less than a hectare) and is within walking distance to nearby stores and restaurants.

WOLF CREEK LODGE 167

UNIT FLOOR PLAN

The residents of Wolf Creek Lodge are a group of independent, active adults who have come together to create a community that supports a cooperative and harmonious way of living. Fostered by patience, open-mindedness, respect, and trust, Wolf Creek Lodge is a place where a strong sense of belonging is shared.

168 CREATING COMMUNITIES

WOLF CREEK LODGE

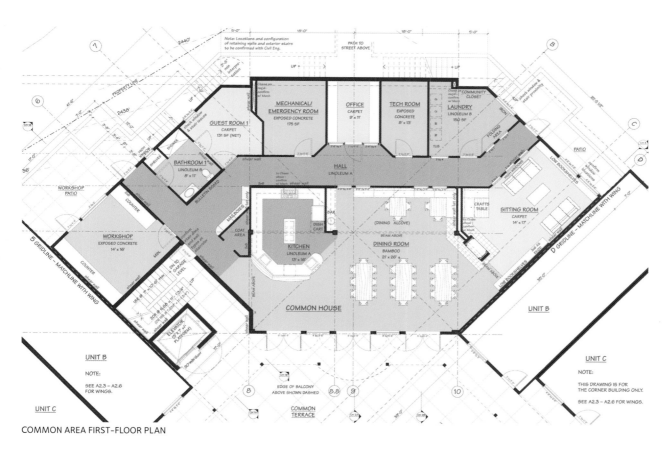

COMMON AREA FIRST-FLOOR PLAN

Rothelowman设计公司

顶点公寓

ROTHELOWMAN

TIP TOP

Location Brunswick East, Victoria, Australia **Units** 411 **Area** 345,392 ft² (32,088 m²)
Photography Scott Burrows (pages 170, 171, 174, 175), Peter Clarke (pages 172 top, 173), Michael Gazzola (page 172 bottom)

In Brunswick East, a new community of 411 residences now occupies the former Tip Top Bakery site, just 2.3 miles (3.7 kilometers) from the Melbourne CBD. Years of planning and community consultation went into this major urban-renewal project neighboring Lygon Street. The design involved the conservation of two important heritage façades—a 1939 administration block and a set of livery stables. Within the master plan, six new low-rise buildings were added, each expressing a unique architectural language, with linear façades inspired by the original bakery's Dutch Modernism, and woven details interpreting aspects of the bread-making process—grain, seeds, rye, malt, silo, and stables.

The façades distil a meaningful response to the site with distinct building identities, and their respective rhythms present a unifying backdrop to the landscaping.

Inside, the apartments are thoughtfully designed and finished. Long-term water and energy use is reduced with passive and active design measures. Rainwater is harvested for public landscape irrigation, while an 8-killowat solar array powers a communal hot water system plus public area and car park lighting, saving approximately 15 tons of greenhouse gas emissions annually.

Diversity and amenity are at the heart of this new community. Housing stock includes entry-level and luxury apartments, townhouses, duplexes, and social housing. A childcare center for residents and neighbors in the Seeds building caters for up to 90 children. To invite the public into this center and the site, a new street, West Lane, was created between Edward and Weston Streets. The internal network of streets and laneways is enriched with landscaped pockets and artwork installations, referencing the site's working past. The basement car park has a single entry/exit point to control traffic and prioritize pedestrian use of laneways, and over 200 bicycle-parking spots are sprinkled throughout the site.

Towards completion of the project in December 2014, the developers, Little Projects, organized the 'Live a Little' festival to bring new and established communities together, celebrating the rebirth of this important site with workshops in bread-making, architectural tours, yoga, childrens' games, and more. In 2014, Tip Top was awarded the UDIA (Victoria) Award for Excellence in Urban Renewal and the API Excellence in Property Award (Heritage), and more recently the 2015 UDIA (National) Award for Urban Renewal.

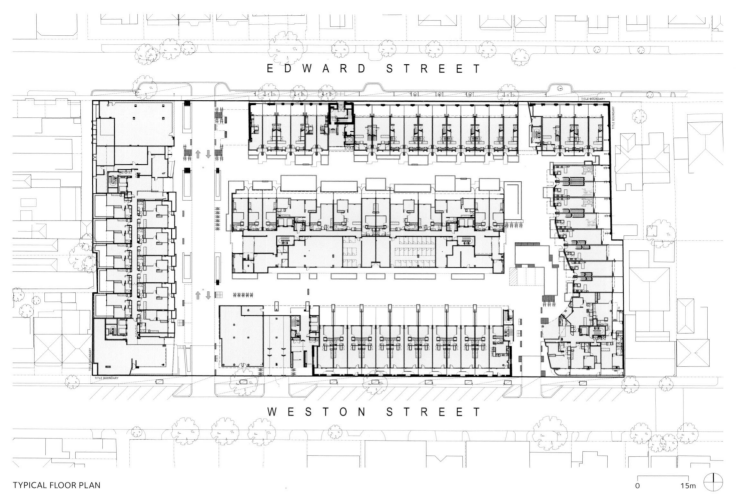

TYPICAL FLOOR PLAN

0　　15m

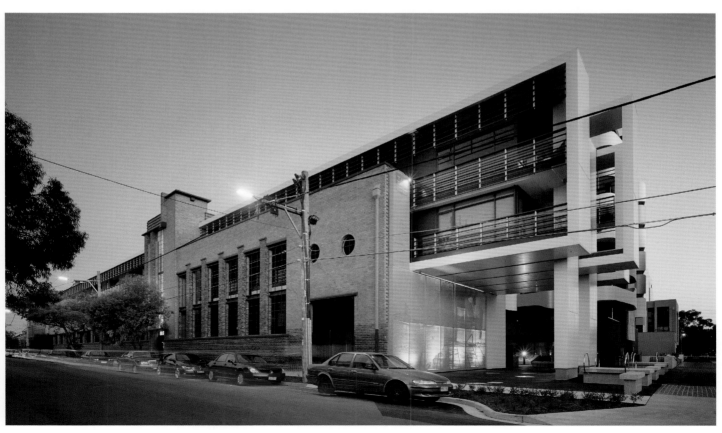

Hyla事务所
新月公寓

HYLA ARCHITECTS
TOH CRESCENT

Location Singapore **Units** 10 **Photography** Derek Swalwell

This cluster housing, residential development consists of ten units of semi-detached houses arranged around an entrance court, which forms a main communal space. The court is dominated by three cascading swimming pools (adult, child, and toddler), with three frangipani trees growing along the water's edge. Situated around the water feature are the unit entrances and perforated granite walls, which enclose each unit's private courtyard. The courtyard and entrance foyer act as a transition zone between public and private spaces.

The living areas face the external perimeter and have their own private garden. On the second level, the junior master faces the central space but is kept private with a timber screen. Adjoining this screen is a balcony, which looks out over the public space, connecting residents to other units and the pool area.

TOH CRESCENT 179

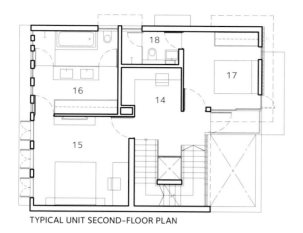

TYPICAL UNIT SECOND-FLOOR PLAN

TYPICAL UNIT ATTIC FLOOR PLAN

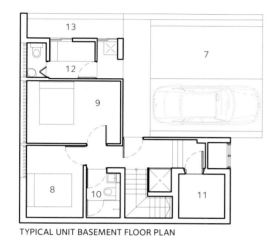

TYPICAL UNIT BASEMENT FLOOR PLAN

TYPICAL UNIT FIRST-FLOOR PLAN

1 FOYER
2 COURTYARD
3 POWDER
4 KITCHEN
5 LIVING / DINING
6 GARDEN
7 CAR PARK
8 BEDROOM 1
9 BEDROOM 2
10 BATHROOM 1
11 HOUSEHOLD SHELTER
12 LAUNDRY / UTILITY / WC
13 YARD
14 STUDY
15 MASTER BEDROOM
16 MASTER BATHROOM
17 BEDROOM 3
18 BATHROOM 2
19 BEDROOM 4
20 BATHROOM 3
21 ROOF TERRACE

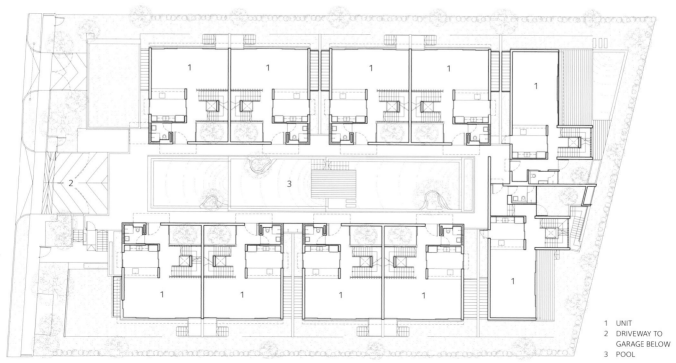

GROUND-FLOOR PLAN

1 UNIT
2 DRIVEWAY TO GARAGE BELOW
3 POOL

The staircase and lift reside in a core-like structure, which has been stepped back from the front of the building, minimizing scale and allowing for improved relation between residence and entrance court. Half the attic has been left as an open roof terrace, further reducing structural scale. Nevertheless, each unit still has five bedrooms, a study and utility room, and two private car spaces.

Internally, the living and dining area form one main space, with an open concept kitchen adjoining it. The staircase has a geometric, aluminum screen and brings in light to all areas of the house. The master bedroom and bathroom face the rear garden in order to maintain privacy.

In typical residential developments, living spaces often face central communal areas, blurring the distinction between private and public space and resulting in a loss of privacy for residents. The Toh Crescent design has sought to properly define and clearly separate public and private space.

贝兹斯玛特设计公司
桶架公寓

BATES SMART
THE GANTRY

Location Sydney, Australia **Units** 191 **Area** 199,132 ft² (18,500 m²) **Photography** Brett Boardman

The Gantry is a residential neighborhood made up of 191 dwellings and located within the inner-city suburb of Camperdown. It consists of four apartment buildings and 26 terraces organized around a large landscaped courtyard. In order to preserve the site's rich history the project integrates new residential buildings with historic industrial building fabric.

A new public laneway and road provide improved site access and mid-block links. Along Australia Street, the sawtooth industrial façades of the 1920s motor car works have been carefully restored and 26 terrace houses are now located within the pitched-roof bays. The gable ends to the historic high-bay building have been retained with party walls and roof, rebuilt to accommodate terrace and loft apartments.

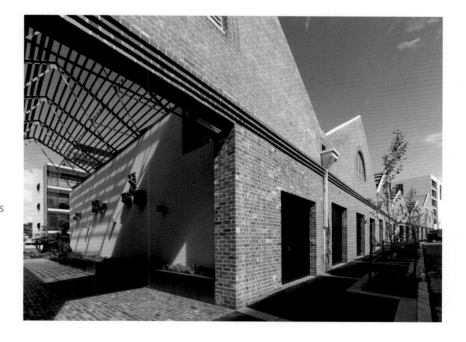

THE GANTRY   185

Three new apartment buildings define the frontage of Denison Street and Parramatta Road. The design of each structure differs in response to its immediate context. Solid masonry façades and louvered screens to Denison Street provide privacy and solar protection to western façades. Balconies to Parramatta Road have operable glass screens to minimize the impact of traffic noise while maintaining solar access and views. Multiple building entries with 'through lobbies' provide visibility between the street and courtyard.

FIRST-FLOOR PLAN

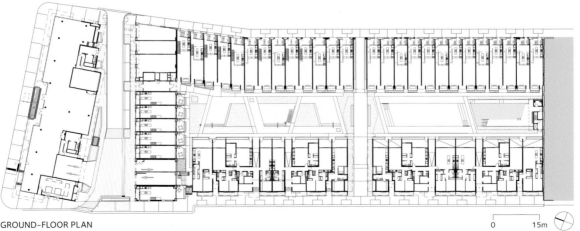
GROUND-FLOOR PLAN

0   15m

创新设计

INNOVATIVE DESIGN

卡达瓦尔和索拉-莫拉莱斯设计公司
科尔多瓦城郊住宅

CADAVAL & SOLÀ-MORALES
CÓRDOBA-REURBANO HOUSING BUILDING

Location Mexico City, Mexico **Units** 9 **Area** 23,680 ft² (2200 m²) **Photography** Miguel de Guzmán

Located in the Colonia Roma neighborhood, in the central sector of Mexico City, ReUrbano is an impressive example of urban regeneration in a region previously devastated by natural disaster.

The project builds on an urban recycling start-up initiative, and the challenge was to harmoniously represent past and present realities while introducing inner-city functionality.

The essence of the project is captured through the preservation of a historically significant residence, which acts as the foundation for the rest of the structure. Built around the remnants of the old site are nine apartments of various sizes and configurations, as well as a commercial area located in the front façade. The existing building occupies the majority of the plot and an outdoor corridor runs down the side of two main cores. Within these are a series of intermediate levels, which construct separate apartments.

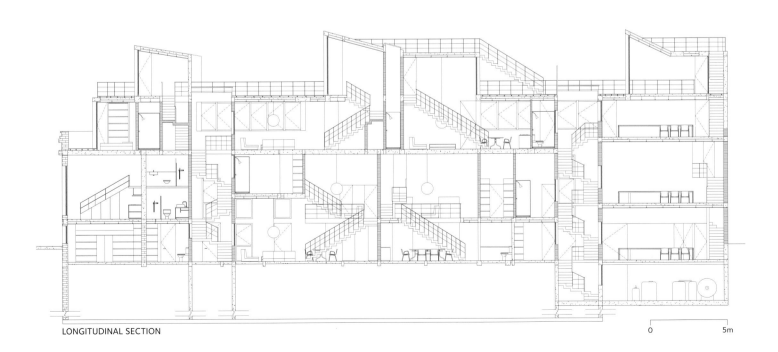

LONGITUDINAL SECTION　　　0　　5m

CÓRDOBA-REURBANO HOUSING BUILDING 193

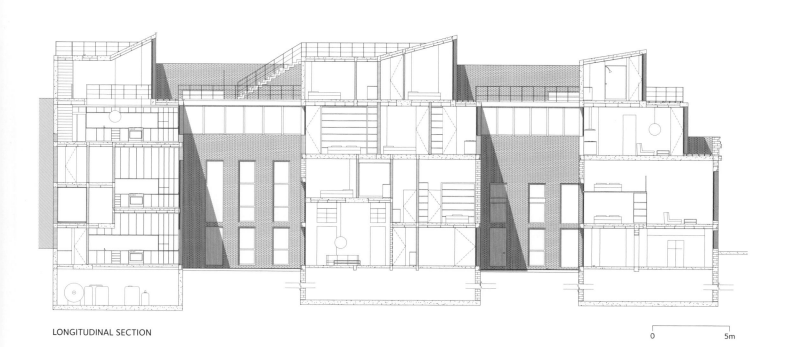

LONGITUDINAL SECTION 0 5m

The interior of the existing house has been transformed from its original single level to up to three levels. The façade of the first floor is built on top of the existing building and is fully glazed to highlight horizontality, lighten the weight of the new addition, and to differentiate the original building from the new intervention. A succession of terraces and built volumes on the top-floor modify the perception of overall building height, and slim the project so it appears as a chain of small towers and not as a continuous solid.

Based on a detailed analysis of existing elements, the overall design aims to generate a new reality built up of two architectural typologies, which respond to two different historical moments. It is a place where the past and present coexist. The successful amalgamation of old and new reconfirms the value of architectural form within the urban grid and its ability to redefine heritage sites.

CÓRDOBA–REURBANO HOUSING BUILDING 195

霍金斯\布朗建筑事务所
立方体公寓

HAWKINS\BROWN
THE CUBE

Location London, United Kingdom **Units** 49 **Area** 72,656 ft² (6750 m²) **Photography** Jack Hobhouse

The Cube has been designed with a twisted cruciform plan, ensuring each apartment has views up and down the neighboring canal basin to the west, or across the park to the east. The cruciform layout turns the traditional inward-looking courtyard block on its head, creating four courtyards that face out toward the surrounding city, providing views and improving access to light and air from apartments that have three external walls. In order to achieve the unique 'twisted-stack' layout, which has floors cantilevering out from the main mass of the building, Hawkins\Brown worked with specialist contractor B+K Structures to develop a cross-laminated timber (CLT) and steel, hybrid structure, built around a reinforced concrete core. The CLT panels and steel-frame elements were all manufactured off site and brought on site for assembly as a kit of parts, reducing the number of trades on site and minimizing construction time. The panels are set into the steel frame, bracing it to form an integral part of the structure. The hybrid structure makes intelligent use of the best properties of both materials to create a lightweight, strong and modern construction, which achieves much lower embodied carbon emissions than an equivalent concrete frame.

THE CUBE 201

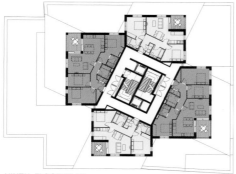
NINTH-FLOOR PLAN

SEVENTH-FLOOR PLAN

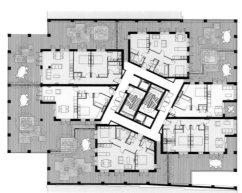
FIRST-FLOOR PLAN

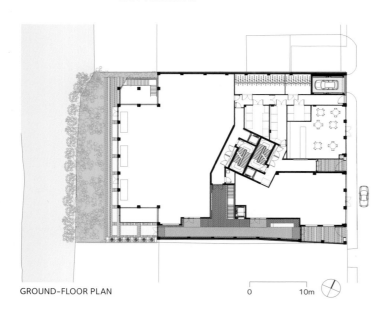
GROUND-FLOOR PLAN

The elevations of the building have been clad in slatted western red cedar. An open screen of black brick creates an orthogonal grid, which wraps around the Wenlock Road elevation of the building, creating visual harmony with neighboring buildings, appropriate to the nearby conservation area.

According to Alex Smith, associate at Hawkins\Brown, 'The Cube was created to be a pioneer of architectural possibility, pushing the boundaries of residential construction and developing homes that are also works of art.'

The 72,656-square-foot (6750-square-meter) scheme is the tallest building to use structural cross-laminated timber in Europe, with its 10 stories reaching a total height of 108 feet (33 meters).

史密斯-米勒建筑事务所与霍金森建筑事务所
狄龙公寓

SMITH-MILLER + HAWKINSON ARCHITECTS
THE DILLON

Location New York, United States **Units** 83 **Area** 176,000 ft² (16,351 m²) **Photography** Michael Moran/OTTO

Due to the structural diversity of units offered, The Dillon provides a unique platform for a variety of city lifestyles. Comprising of 83 residential units, ranging from studios to five-bedroom apartments, 57 of them have been designed with unique floor plans.

The project's dense city-living fabric has been achieved by joining maisonette, duplex, triplex, and studio apartments in a cosmopolitan weave—a configuration that pays homage to both the ubiquitous New York City brownstone block and Le Corbusier's *Ville Radieus*, as well as descendants such as Gordon Bunshaft's *Manhattan House*, James Freed's *Kip's Bay*, and Oscar Stonoroff's *NYU Housing*.

As an extensive low-rise and mid-block project, the building replaces open parking lots and derelict structures with an optimistic premise drawn, in part, from Jane Jacobs' observations of city life.

THE DILLON   207

In an innovative response to restrictive zoning, the through-block site hosts both the customary residential tower, with repeating floor plans, and an a-typical model; a bar building combining maisonettes, skip-stop duplex and skip-stop triplexes, with roof-top cabanas all served by underground self-parking facilities. The assemblage of these elements results in a building section of remarkable economy and presents a new typology for urban living.

The south-facing street façade was designed as a high performance curtain wall to provide access to daylight, ventilation, and to maximize views and connection to the city. The north-facing court was designed around ground-floor garden court spaces. These views are carefully modulated utilizing low-E coatings and lenticular films to minimize solar gain and create privacy from the street. Internal solar shades and projected sash windows give inhabitants the ability to further modulate the environmental quality inside the units. The parallel projected sash units ventilate up and down simultaneously, increasing the efficiency and depth at which fresh air is circulated within the unit.

The project has an informal quality that appears to step down the street. Its unique folded façade offers views across and along the street, westward to the Hudson River, and eastward for early morning light. All units optimize the available light and view, and many enjoy both southern street and northern court exposure with natural cross ventilation. The maisonettes' small street-side forecourts buffer domestic activity while affording direct street access.

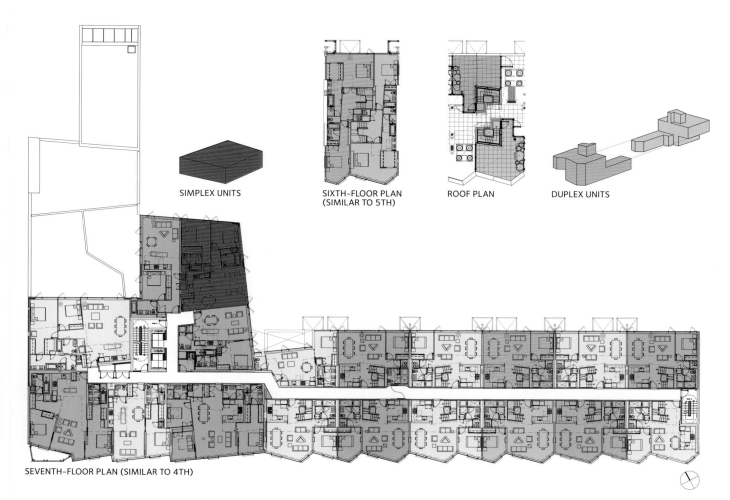

SIMPLEX UNITS   SIXTH-FLOOR PLAN (SIMILAR TO 5TH)   ROOF PLAN   DUPLEX UNITS

SEVENTH-FLOOR PLAN (SIMILAR TO 4TH)

托尼·欧文公司
阿尔法公寓

TONY OWEN PARTNERS
THE ALPHA

Location Lewisham, New South Wales, Australia **Units** 70 **Area** 70,535 ft² (6553 m²) **Photography** Steve Back

The key intention behind the unique design of The Alpha was to demonstrate that progressive developmental techniques are as financially viable as traditional architectural designs; a notion that Tony Owen Partners admit was the greatest challenge with this project. Each of the 70 one- and two-bedroom units within The Alpha is afforded a unique inner-city relationship with the exterior environment by creating simultaneous privacy from and exposure to the surrounding Sydney suburbs through the perspective of the distinctive, hexagonal pod-like façade. The construction of the hexagonal façade was highly efficient, requiring only a single aluminium layering, which could be combined with a nodal connection, amounting to only 2.5 percent of construction costs.

212   INNOVATIVE DESIGN

THE ALPHA 213

FIRST-FLOOR PLAN

THIRD-FLOOR PLAN

The volume stretches seven stories and incorporates a retail space on the ground floor. The three-bedroom units are reserved for the top-story 'skyhomes,' whose hexagonal pods open onto an extensive roof terrace, offering wide views toward the skyline of Sydney and the luscious green landscape of Lewisham. These skyhomes have two-story living areas, loft bedrooms, and dramatic stair voids, with the upper level opening onto a private rooftop garden for each unit. Profiled concrete and coloured glass are utilised to accentuate the bold lobbies of the building, while the ground floor encompasses the childcare facilities. Each residential level also has a unique two-story communal lounge to encourage interaction between residents.

The Alpha forms the new cornerstone of the McGill design precinct—a stylish, mixed-use designer precinct with contemporary buildings and public open spaces—and represents a futurist characterization of urban architecture in Sydney. The vision for this area combines the heritage values and rich history of Lewisham with modern, ecologically sustainable development.

柯林斯·加达耶建筑事务所
汉普顿环道公寓与联排别墅

COLLINS CADDAYE ARCHITECTS
HAMPTON CIRCUIT APARTMENTS & TOWNHOUSES

Location Canberra, Australia **Photography** Stefan Postles

The Hampton Circuit Apartments & Townhouses are the outcome of competing aspirations—the client's ambitious brief for a unique and memorable design, and a process of architectural enquiry that sought to fulfill the client's vision while also providing a fit-for-purpose multihousing development.

The process of discovering what the building should look and feel like was protracted and required a setting aside of preconceptions. A looseness and freedom-of-hand was used to design an expressive building and satisfy the client's brief for a distinctive form. However, behind the seemingly free-form design is a logic that addresses the functional requirements of multihousing typology.

The form of the rear five-story apartment building is a result of being 'pulled' forward to provide address to a constrained street frontage on Hampton Circuit. Conversely, the apartment building gently turns its back to the noise and disruption of Adelaide Avenue, providing a protective sense of enclosure to the central landscaped areas.

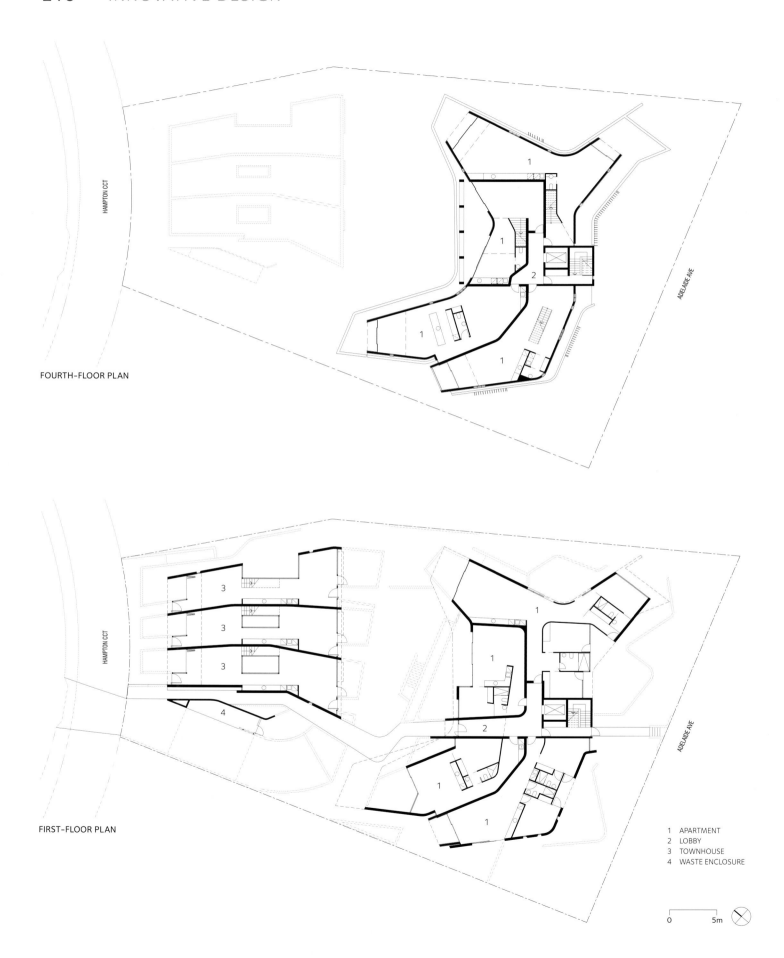

HAMPTON CIRCUIT APARTMENTS & TOWNHOUSES

Spaces stretch out to capture the sun and encase long-range views of the National Arboretum, Black Mountain and The Lodge. The generous distance between the townhouses at the front of the site and the apartment building behind allow for unimpeded solar access. West- and south-facing windows have been minimized, both as an energy saving strategy and in response to the noise generated from the street. Due to the fluid nature of the building forms, spaces are continuous and able to open and close as required, optimizing natural ventilation.

A limited external materials palette is a deliberate counter-point to the seemingly complex forms. It comprises two primary materials—off-form gray concrete and black lightweight cladding—with the robust and industrial nature of these materials imparting a sense of restraint and order. Internal spaces are free-flowing and flexible, however, the organic quality is underpinned by considered logic regarding the placement of furniture and fittings, allowing occupants to comfortably inhabit the spaces in a multitude of ways.

Dekleva Gregorič建筑事务所
布里克社区

DEKLEVA GREGORIČ ARCHITECTS
BRICK NEIGHBORHOOD

Location Ljubljana, Slovenia **Units** 185 **Area** 217,269 ft² (20,185 m²) **Photography** Miran Kambic

Brick Neighborhood's development and design principles were underpinned by a goal to establish a clear spatial, material, and social neighborhood identity. Architect Dekleva Gregoric's overall aim was to create a deep sense of connection between residents and their living environment.

The selection of bricks as a preliminary material comes from a memory of the brickyard that used to be on site. They also provide an opportunity for further architectural expression through the manipulation of material and micro-structuring of the façade surface, ensuring originality within the community.

Brick Neighborhood consists of 185 individual dwellings, made up of 17 diverse flat types, which address the various lifestyle needs of residents.

The position of the structure, installation and internal organisation of the flats revolve around a backbone service strip, which enables internal flexibility with a diverse set of rooms distributed either as one large unified space, or a set of smaller rooms. The system allows the joining of smaller flats or separating of larger units, before, during, and after construction.

BRICK NEIGHBORHOOD 225

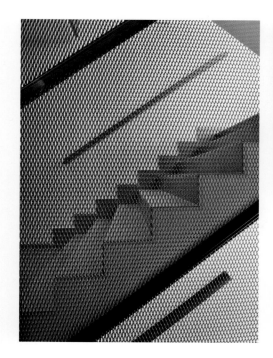

Communications are naturally illuminated from two sides and allow the corridor to act as a place of meeting. Social interaction is upgraded by placing a shared common space above the entrance of each building, providing the possibility for birthday parties, indoor playgrounds for wintery months, gym, and other activities.

On the glazing within each communal space, Slovenian poems have been written. This contextual and semantic upgrade of the architecture enriches the cultural awareness of residents and allows them to further identify with their living environment.

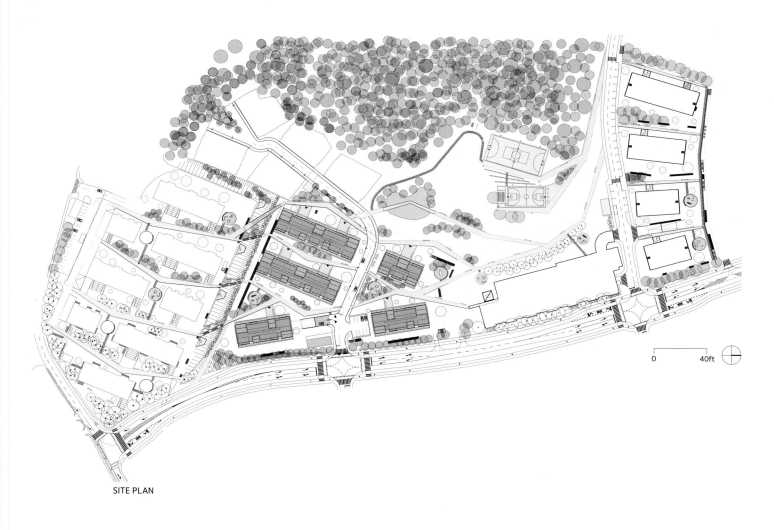

SITE PLAN

杰克逊·克莱门茨·巴罗斯建筑事务所
上院公寓

JACKSON CLEMENTS BURROWS ARCHITECTS
UPPER HOUSE

Location Melbourne, Australia **Units** 110 **Photography** John Gollings (pages 226, 227), Shannon McGrath (pages 228, 229, 230, 231)

The multi-award-winning Upper House demonstrates clarity in urban composition and a social program that engages with its context in a new and positive way. Located in Melbourne, Australia, the 17-story building consists of 110 apartments and two commercial tenancies.

Upper House features a unique composition: its communal space was imagined at the heart of the building, with upper levels floating above like clouds. Central to the project as an amenity, the space also demonstrates principles that acknowledge sustainability as both social and environmental.

Apartments are decorated in restful neutrals, allowing visual impact and accents to be tailored to each resident's own style. The communal space is elegantly furnished and provides for a range of day- and night-time activities.

GROUND-FLOOR PLAN

1 COMMERCIAL TENANCY 1
2 COMMERCIAL TENANCY 2
3 MAIN ENTRY FOYER
4 BIKES
5 BIN ROOM
6 PUMP ROOM
7 SUBSTATION
8 FIRE CONTROL ROOM
9 DISABLED TOILET

TYPICAL PODIUM FLOOR PLAN (LEVELS 1–10)

UPPER HOUSE 231

An important consideration and compositional design outcome was the designation of level 11 as a communal space for the residents. It incorporates a lounge, gymnasium, and dining space overlooking a green roof scape.

The building's podium is a solid form of natural concrete and contrasts with the partially transparent, white glass curtain wall of the upper form. This contrast amplifies a sense of formal break in the massing, realizing the vision for a building that floats across the skyline 'like a cloud'. The glass finish enables a degree of reflectivity and suggests a more ephemeral lightweight gloss surface, which is offset against the matte concrete base. The mirrored soffit of the upper form creates reflections of the gardens and social activity, providing a visual cue to the elevated garden from street level.

Across the upper and lower façades are a scattering of balconies and windows. Varying cantilevered elements and alternating window placements produce an engaging three-dimensional quality, creating a façade with its own topography.

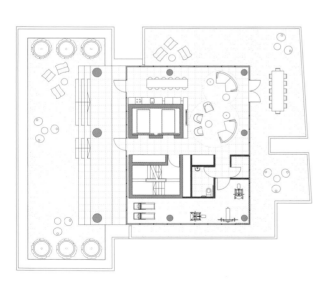

OBSERVATORY FLOOR PLAN (LEVEL 11)

TYPICAL 'CLOUD' FLOOR PLAN (LEVELS 12–16)

0   5m

A-lab设计公司
雕刻公寓

A-LAB
THE CARVE

Location Oslo, Norway **Units** 41 **Area** 242,188 ft² (22,500 m²) **Photography** Luis Fonsica, Ivan Brodey

The Carve is an non-traditional, high-rise apartment building enveloping a narrow strip of 69 feet by 177 feet (21 meters by 105 meters), with a maximum-height regulation of 177 feet (54 meters). The white marble and wood-panel clad building embodies a mix-use complex totaling 15 stories.

The first eight floors are designated office space with a residential program and total 236,806 square feet (22,000 square meters). The mixed program is structured by compacting the flexible office spaces to create efficiency, while optimizing views and outdoor spaces of the apartments situated around a raised, undercover garden.

A public passage cuts through the first two levels, facilitating a pedestrian route through all barcode buildings, while also generating space for a separate entrance to the residential floors, connecting them directly to Oslo Central Station.

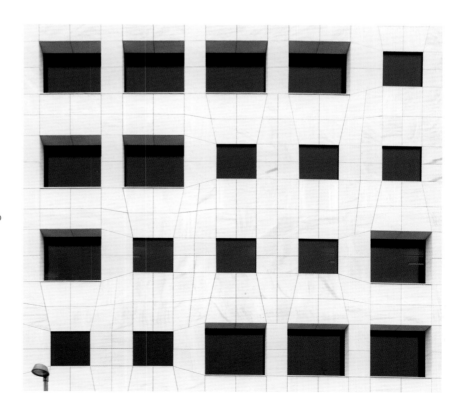

236 INNOVATIVE DESIGN

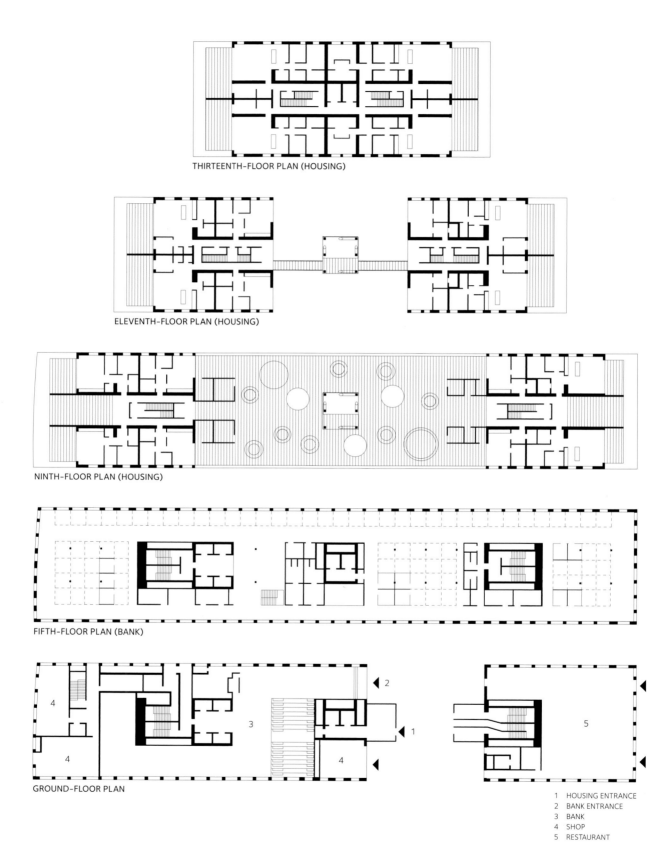

THIRTEENTH-FLOOR PLAN (HOUSING)

ELEVENTH-FLOOR PLAN (HOUSING)

NINTH-FLOOR PLAN (HOUSING)

FIFTH-FLOOR PLAN (BANK)

GROUND-FLOOR PLAN

1 HOUSING ENTRANCE
2 BANK ENTRANCE
3 BANK
4 SHOP
5 RESTAURANT

THE CARVE 237

The project is situated in the Opera Quarter of the new Oslo waterfront development and offers idyllic views of Oslo Fjord and Oslomarka.

Only five minutes from the central station and airport fast-train, location is one of the project's best assets.

The residential complex rests on 10,764 square feet (1000 square meters) of common open area, which includes an elevated garden terrace, creating distance from the corporate world beneath. Fitted with a panoramic elevator and open-air bridges, the green foyer acts as a buffer zone for residents passing through on their way home. Both ends open onto communal terraces overlooking the Oslo Fjord to the south, and cityscapes to the north and east.

The covered garden is a physical response to the uncompromising real estate strategy of the barcode plan where, in order to raise the environmental standards and living quality in the new city development, one-sided apartments are not allowed. The result is equally uncompromising; one architectural gesture removes the one-side apartments and creates a roof garden to two new inner façades. This is one of the structure's most notable features and allows for a terraced-housing typology in the center of Oslo.

索引
INDEX OF ARCHITECTS

A-lab
www.a-lab.no
The Carve 232–7

Aytac Architects
www.aytacarchitects.com
Apartman 18 108–113

Bates Smart
www.batessmart.com
The Gantry 182–7

Benjamin Hall Design
www.benjamin-hall-p2w2.squarespace.com
White Stone Studios 126–31

bureau^proberts
www.bureauproberts.com.au
SILT Apartments 132–7

Burnazzi Feltrin Architects
www.burnazzi-feltrin.it/en/
GI multi-family housing 26–31

Cadaval & Solà-Morales
www.ca-so.com
Córdoba-ReUrbano Housing Building 190–5

Casanova+Hernandez Architects
www.casanova-hernandez.com
Black & White Twins 100–5

Collins Caddaye Architects
www.collinscaddaye.com.au
Hampton Circuit Apartments & Townhouses 214–19

David Baker Architects
www.dbarchitect.com
Richardson Apartments 70–5

Dekleva Gregorič Architects
www.dekleva-gregoric.com
Brick Neighborhood 220–5

Designgroup Stapleton Elliott
www.designgroupstapletonelliott.co.nz
Marshall Court 152–7

Elenberg Fraser
www.elenbergfraser.com
Abode318 138–43

GBD Architects
www.gbdarchitects.com
Kiln Apartments 50–5

Guillermo Vázquez Consuegra Architect
www.vazquezconsuegra.com
Social Housing in Vallecas 88–93

Hawkins\Brown
www.hawkinsbrown.com
The Cube 196–201

Helen & Hard
www.helenhard.no
Rundeskogen 20–5

Hyla Architects
www.hyla.com.sg
Toh Crescent 176–81

Jackson Clements Burrows Architects
www.jcba.com.au
Upper House 226–31

Jonathan Segal FAIA
www.jonathansegalarchitect.com
The North Parker 82–7

Justin Mallia
www.justinmallia.com
Yan Lane 44–9

Kevin Daly Architects
www.kevindalyarchitects.com
Broadway Housing 76–81

KieranTimberlake
www.kierantimberlake.com
Belles Townhomes 38–43

LAPS Architecture
www.laps-a.com
Civic Center + Social Housing 58–63

MAArchitects
www.maarchitects.com.au
Walsh Street 144–9

MAB Arquitectura
www.mabarquitectura.com
Civic Center + Social Housing 58–63

McCamant & Durrett Architects
www.cohousingco.com
Wolf Creek Lodge 164–9

Neometro
www.neometro.com.au
Walsh Street 144–9

Peter Pichler Architecture
www.peterpichler.eu
The Mirror Houses 120–5

rh+ architecture
www.rhplus-architecture.com
Plein Soleil 64–9

Rothelowman
www.rothelowmaninsight.com
Tip Top 170–5

Smith-Miller + Hawkinson Architects
www.smharch.com
The Dillon 202–7

Steinsvik Architecture
www.steinsvikarkitekt.no
Passivhus Storelva 14–19

Studio Libeskind
www.libeskind.com
Vitra 114–19

Teeple Architects
www.teeplearch.com
60 Richmond Street 94–9

Tony Owen Partners
www.tonyowen.com.au
The Alpha 208–13

YUUA Architects & Associates
www.yuua.jp
NOIE Cooperative House 158–63

Every effort has been made to trace the original source of copyright material contained in this book. The publishers would be pleased to hear from copyright holders to rectify any errors or omissions.

The information and illustrations in this publication have been prepared and supplied by Avi Friedman and the contributors. While all reasonable efforts have been made to ensure accuracy, the publishers do not, under any circumstances, accept responsibility for errors, omissions and representations express or implied.